Mathematical Problem Solving

Workbook 5

Strategy for Solving Real-World Problems

Satya Pradhan

ISBN: 1541176081

ISBN 13: 9781541176089

Library of Congress Control Number: 2016921389

CreateSpace Independent Publishing Platform, North Charleston, SC

Contents

Introduction

Having strong problem-solving skills can make a huge difference in one's career in the modern knowledge-based economy. Problems are at the center of what we do at work every day. Whether one is developing a vaccine for the flu, creating antivirus programs for the Internet, delivering lifesaving drugs to remote villages, maximizing profits for a company, or understanding the complex structure of our universe, problems are an integral part of our everyday lives. So being an effective and confident problem solver is really important to one's success. Much of that confidence comes from having a good understanding of strategy and the tools to use when approaching a problem. Therefore, it is essential for students to develop the skills and techniques for problem solving from an early age, when they are in elementary school.

Conceptual understanding, procedural and computational skills, and application of concepts to real-life problems are three pillars of mathematics education. Conceptual understanding involves knowing what to do, procedural fluency requires knowing how to do it, and problem solving focuses on solving a wide variety of complex, real-life problems using mathematical knowledge. Mathematical skills have been taught in school in this order: maximum emphasis is placed on the understanding of math concepts and computational skills, followed by the application of the concepts to real-world problems. However, real-life problem solving requires students to apply these concepts in the exact opposite order, starting with understanding the problem, then finding the mathematical concepts required to solve the problem, and finally choosing the method that best solves the problem.

Mathematical problem solving is often taught as a way to reinforce mathematical concepts, which misses the importance of strategic thinking while solving a problem. Many research articles and books have been written emphasizing the importance of problem-solving strategy. However, the burden of teaching problem-solving strategy is left mostly to teachers and parents, who are expected to develop their own curriculum and lesson plans for the complex topic of strategy and then teach it to students.

This book presents several problem-solving strategies that can easily be used by teachers and parents to teach the subject. The first two chapters present the concepts of number operations and the basic problem-solving strategies listed below:
- Solving one-step problems
- Solving multistep problems
- Solving problems working backward
- Formulating problems with variables and equations
- Solving problems using variables

The concept of the unitary method is presented in chapter 3. The remaining chapters present lessons on different problem types. The objective is to teach students how to start with a problem statement, understand the problem, and then solve it with a known mathematical procedure.

There are many different problem types that students will encounter in their careers. The following problem types have been selected because they are appropriate for students in fifth grade:
- Number problems
- Age problems
- Time and distance problems
- Money problems
- Work problems
- Mixture problems

Each lesson in the workbook is classified as (*), (**), or (***), depending on the level of difficulty, and each starts with a few examples showing how to solve a particular type of problem. This is followed by a number of problems of this type.

Notes to Parents, Teachers, and Tutors

As a parent, you can use this workbook to teach problem-solving techniques to your child without any teaching experience. The first three chapters present the basic concepts and should be taught first. If students are already familiar with these concepts, these chapters can be skipped. All other chapters are independent of one another and can be taught in any order.

As a teacher, you can easily integrate this workbook into the school curriculum by choosing appropriate lessons to be taught along with the curriculum.

Private tutors and after-school learning centers can use this workbook to offer special classes on mathematical problem solving or as part of other math enrichment programs. Our suggestion is to teach two or three lessons a week, using the example questions given at the beginning of each lesson and giving the other questions as homework.

Conceptual understanding of mathematical problem solving is the main focus of this book. Therefore, students are encouraged to use a calculator to solve the numerical expressions. This will allow them to take less time for numerical calculation and focus on understanding concepts.

Online Practice Lessons (www.ilecy.com)

Teachers, parents, and tutors can give the online lessons available on the website www.ilecy.com as homework. The online practice lessons have many advantages, such as detailed solutions for each question, automatic grading and feedback when the student completes a question, tracking of students' daily usage, and more. Please visit www.ilecy.com to learn more about the online lessons and how to use them in your classroom.

Answer Keys

Answer keys for all questions in this book are available online. You can download the PDF file for the answer keys from www.ilecy.com/AnswerBooks.

Feedback

Feedback is always welcome from students, parents, and teachers. Please send your comments, testimonials, or suggestions for improvement to mathPS100@gmail.com.

Acknowledgments

My sincere appreciation and thanks goes out to the following people for their feedback and suggestions while teaching these lessons as part of an after-school math enrichment program: Susie Bierman, Reynaldo Lorenzana, Shalini Kinger, Arun Sahoo, and Alicia Lopez. I would also like to acknowledge the help of Soumya Sahoo, Kallala Giri, Arpita Rout, and Subhashis Dash during the preparation of this workbook. My sincere appreciation to my wife, Nishi, for her support while I was working on this book along with my hectic, full-time job in Silicon Valley, CA. Special thanks to my son, Sougat, and daughter, Sarika, for their valuable feedback on different lessons. They were my first reviewers, connecting the lessons to their classroom in school.

Assessment

Note: Some of the assessment questions may be challenging at the beginning fifth grade.

Chapter-1:

1. What operation will you use for the key word *loss*?
   (a) Addition
   (b) Division
   (c) Subtraction
   (d) Multiplication

   Answer: _____

2. What is the math sentence for the following expression?

   Total of 23 and 24 candies are distributed equally among 9 kids
   (a) $(23 + 24) \div 9$
   (b) $(23 \times 24) - 9$
   (c) $(23 \div 24) + 9$
   (d) $(23 + 24) \times 9$

   Answer: _____

Chapter-2:

3. Mike can pick 85 fruits in an hour. How many fruits can he pick in 36 minutes?

   Answer: _____

4. Angela bought some ribbons for $0.40 each, pens for $3.00 each, and soap boxes for $22.00. If the cashier returned $4.60, how much money did Angela give to the cashier?

   Answer: _____

5. The sum of Ria and her dad's ages is 52. Ria's dad is 30 years older than Ria. Use the variable M to represent Ria's age. Which equation represents the given problem?
   (a) $M + 30 = 52$
   (b) $M - 30 = 52$
   (c) $2M + 30 = 52$
   (d) $2M - 30 = 52$

   Answer: _____

Chapter-3:

6. If 4 buffalos can produce 80 liters of milk, how many buffalos will produce 120 liters of milk?

   Answer: _____

7. 12 plumbers can complete a task in 6 hours. How long will 8 plumbers take to complete the same task?

   Answer: _____

8. Kiran makes 32 toys in 20 hours. How many toys will he make in 15 hours?

   Answer: _____

Chapter-4:

9. What is the sum of the values of 4, 9, and 8 in 45,968?

    Answer: _____

10. What is the largest 3-digit number that is divisible by 10?

    Answer: _____

11. Find the GCF of the following numbers.

    | 10, 25, 30 ____

    Answer: _____

12. What is the smallest number that must be added to 100 so the sum is divisible by 24, 40, and 60?

    Answer: _____

13. Find the LCM of the following numbers.

    | 35, 60, 75 ____

    Answer: _____

Chapter-5:

14. Jimmy is currently 13 years old. How old will he be 8 years from now?

    Answer: _____

15. The difference between John's and Rob's ages is 10. What was the difference between their ages 7 years ago?

    Answer: _____

16. Roma is currently 2 times as old as Nikhil. The difference between their ages is 6. What is Nikhil's age?

    Answer: _____

17. In April of 2002, Sofia was 2 times as old as her brother. The sum of their ages was 12. What was her brother's age in April 2002?

    Answer: _____

18. Soham is presently 7 years older than Alex. The sum of their present ages is 35. How old will Soham be in 5 years?

    Answer: _____

Chapter-6:

19. Find the unit rate for the following expression:

    48 cookies for 6 children

    Answer: _____

20. David and his friends traveled for 6 hours. They traveled at a speed of 75 miles per hour for the first 3 hours and at 65 miles per hour for rest of the trip. How far did they travel?

    Answer: _____

21. Mrs. Rao walks for 600 meters every morning. Her average speed is 25 meters per minute. How long does she walk every morning?

    Answer: _____

22. One bag contains 8 fruits and can serve 4 children. Find how many bags are needed to serve 16 children.

    Answer: _____

23. If a racing car can travel 320 kilometers in 1 hour and 20 minutes, what is the average speed in kilometers per minute?

    Answer: _____

Chapter-7:

24. Roni deposited $6,000.00 in her bank account that will earn a simple annual interest rate of 7 percent for a period of 48 months. What is the time in years?

    Answer: _____

25. Jack deposited $5,000.00 in his bank account. The bank will give him 6 percent annual interest. If he has to withdraw the balance after 5 years, how much will he withdraw?

    Answer: _____

26. A principal amount of $2,500.00 earns simple interest and becomes $2,750.00 in 2 years. What is the simple annual interest rate?

    Answer: _____

27. Use the simple interest formula to find the unknown quantity.

    P = $450.00
    R = 7 percent
    t = 2 years
    I = ?

    Answer: _____

28. Amit is investing in a mutual fund. He will be getting a simple interest rate of 8 percent per year. How much money should he invest to earn $1,400.00 interest in 5 years?

    Answer: _____

Chapter-8:

29. It takes 9 engineers to complete 3 surveys in a weak. How many engineers are needed to complete 4 surveys in a weak?

Answer: _____

30. Bill takes 2 hours to make a wall painting. Bob takes 3 hours to make the same wall painting. If Bill and Bob work together, how long will they take to make the painting?

Answer: _____

31. It takes 8 tailors 12 hours to sew some clothes. How many hours will it take 12 tailors to sew the same clothes?

Answer: _____

32. Two people can wash 4 cars in an hour. How many people are needed to wash 6 cars in the same time?

Answer: _____

Chapter-9:

33. A container has 40 liters of milk. If we add 10 liters of water to the container, what will be the percentage of milk in the solution?

Answer: _____

34. Two bags have 50 marbels each. Bag 1 has 50 percent red marbel, and bag 2 has 30 percent red marbel. How many more red marbels are in bag 1 than in bag 2?

Answer: _____

35. Box A has 30 red M&M and 20 blue M&M. Box B has 20 red M&M and 30 blue M&M. If we mix the M&Ms from both the boxes, what will be percentage of red M&M in the mixture?

Answer: _____

36. Pipe 1 can fill a swimming pool in 8 hours. Pipe 2 can fill the same pool in 12 hours. If both the pipes are opened at the same time when the pool is empty, how long will it take to fill the pool?

Answer: _____

1. Mathematical Operations

1.1 Addition and Subtraction Key Words (*)

Example 1:

What operation will you use for the key word *gain*?
  (a) Subtraction
  (b) Addition
  (c) None of the above

Solution:

The key word *gain* is an addition key word. So the answer is (b).

Example 2:

What is the addition or subtraction key word(s) in the following phrase?

Difference between 15 and 6

Solution:

In this expression, *difference* and *between* are the subtraction key words.

Write or choose the letter of the answer.

1. What operation will you use for the key words *sum of*?
  (a) Addition
  (b) Subtraction
  (c) Both (a) and (b)
  (d) None of the above

Answer: _____

2. What is the operation key word(s) in the following sentence?

John has 3 more pens than Andy.

Answer: _____

3. What operation will you use for the key word(s) in question 2?
  (a) Subtraction
  (b) Addition
  (c) Both (a) and (b)
  (d) None of the above

Answer: _____

4. What is the operation key word(s) in the following phrase?

Total of 50 and 60

Answer: _____

5. What operation will you use for the key words *decreased by*?
  (a) Addition
  (b) Subtraction
  (c) Both (a) and (b)
  (d) None of the above

Answer: _____

6. What operation will you use for the key word *more*?
  (a) Addition
  (b) Subtraction
  (c) Both (a) and (b)
  (d) None of the above

Answer: _____

Write or choose the letter of the answer.

7. What is the operation key word(s) in the following sentence?

    The sum of 102 and 712 is equal to 814.

    Answer: _____

8. What operation will you use for the key word(s) in question 7?
    (a) Addition
    (b) Subtraction
    (c) Both (a) and (b)
    (d) None of the above

    Answer: _____

9. What is the operation key word(s) in the following sentence?

    Find the difference between 41 and 23.

    Answer: _____

10. What operation will you use for the key word(s) in question 9?
    (a) Addition
    (b) Subtraction
    (c) Both (a) and (b)
    (d) None of the above

    Answer: _____

11. What operation will you use for the key word *minus*?
    (a) Addition
    (b) Subtraction
    (c) Both (a) and (b)
    (d) None of the above

    Answer: _____

12. What operation will you use for the key word *plus*?
    (a) Addition
    (b) Subtraction
    (c) Both (a) and (b)
    (d) None of the above

    Answer: _____

13. What operation will you use for the key word *fewer*?
    (a) Addition
    (b) Subtraction
    (c) Both (a) and (b)
    (d) None of the above

    Answer: _____

14. What is the operation key word(s) in the following expression?

    12 plus 15

    Answer: _____

15. What operation will you use for the key words *raise of*?
    (a) Addition
    (b) Subtraction
    (c) Both (a) and (b)
    (d) None of the above

    Answer: _____

16. What is the operation key word(s) in the following sentence?

    Kapil had a gain of $1,300.00 last year.

    Answer: _____

1.2 Multiplication and Division Key Words (*)

Example 1:

What operation will you use for the key word *per*?

(a) Multiplication
(b) Division
(c) None of the above

Solution:

Per is a division key word. So the answer is (b).

Example 2:

What is the multiplication or division key word(s) in the following sentence?

Find the double of 1.8.

Solution:

In this sentence, *double* is the multiplication key word.

Write or choose the letter of the answer.

1. What is the operation key word(s) in the following sentence?

Find the average speed of 25 miles per 5 hours.

Answer: _____

2. What operation will you use for the key word(s) in question 1?

(a) Multiplication
(b) Division
(c) None of the above

Answer: _____

3. What operation will you use for the key word *times*?

(a) Multiplication
(b) Division
(c) None of the above

Answer: _____

4. What operation will you use for the key words *divided equally*?

(a) Division
(b) Multiplication
(c) None of the above

Answer: _____

5. What operation will you use for the key word *multiply*?

(a) Multiplication
(b) Division
(c) None of the above

Answer: _____

6. What is the operation key word(s) in the following sentence?

Ayan drinks coffee twice a day.

Answer: _____

Write or choose the letter of the answer.

7. What is the operation key word(s) in the following sentence?

    Find the quotient of 20 and 5.

    Answer: _____

8. What is the operation key word(s) in the following sentence?

    Alice saves 25 percent of her salary.

    Answer: _____

9. What is the operation key word(s) in the following sentence?

    One-fourth of 20 tomatoes are rotten.

    Answer: _____

10. What operation will you use for the key word(s) in question 9?
    (a) Division
    (b) Multiplication
    (c) None of the above

    Answer: _____

11. What is the operation key word(s) in the following sentence?

    Divide 33 by 11.

    Answer: _____

12. What is the operation key word(s) in the following sentence?

    The 42 toy cars are divided equally among 7 kids.

    Answer: _____

13. What operation will you use for the key word(s) in question 12?
    (a) Division
    (b) Multiplication
    (c) None of the above

    Answer: _____

14. What operation will you use for the key words *percentage of*?
    (a) Multiplication
    (b) Division
    (c) None of the above

    Answer: _____

15. What is the operation key word(s) in the following sentence?

    Sam went for the exam 3 times every week.

    Answer: _____

16. What operation will you use for the key words *product of*?
    (a) Multiplication
    (b) Division
    (c) None of the above

    Answer: _____

17. What operation will you use for the key words *quotient of*?
    (a) Multiplication
    (b) Division
    (c) None of the above

    Answer: _____

1.3 Operation Key Words 1 (*)

Example 1:

What is the operation key word(s) in the following sentence?

Divide 315 by two-thirds.

Solution:

In this expression, *divide by* are the division key words.

Example 2:

What operation will you use for the key word *loss*?
- (a) Addition
- (b) Multiplication
- (c) Subtraction
- (d) Division

Solution:

The key word *loss* is a subtraction key word. So the answer is (c).

Write or choose the letter of the answer.

1. What is the operation key word(s) in the following expression?

   Sum of 5 and 17

   Answer: _____

2. What operation will you use for the key word(s) in question 1?
   - (a) Addition
   - (b) Subtraction
   - (c) Multiplication
   - (d) Division

   Answer: _____

3. What is the operation key word(s) in the following sentence?

   Last year John had a loss of $1,200.00 in his business.

   Answer: _____

4. What operation will you use for the key word(s) in question 3?
   - (a) Division
   - (b) Subtraction
   - (c) Multiplication
   - (d) Addition

   Answer: _____

5. What operation will you use for the key word *remainder*?
   - (a) Subtraction
   - (b) Multiplication
   - (c) Addition
   - (d) Division

   Answer: _____

6. What operation will you use for the key word *times*?
   - (a) Division
   - (b) Addition
   - (c) Subtraction
   - (d) Multiplication

   Answer: _____

Write or choose the letter of the answer.

7. What is the operation key word(s) in the following expression?

    10 centimeters of ruler tripled

    Answer: _____

8. What operation will you use for the key words *decreased by*?
    (a) Subtraction
    (b) Division
    (c) Addition
    (d) Multiplication

    Answer: _____

9. What is the operation key word(s) in the following expression?

    96 miles per 3 hours

    Answer: _____

10. What operation will you use for the key word(s) in question 9?
    (a) Addition
    (b) Multiplication
    (c) Division
    (d) Subtraction

    Answer: _____

11. What operation will you use for the key words *divided equally*?
    (a) Division
    (b) Subtraction
    (c) Addition
    (d) Multiplication

    Answer: _____

12. What operation will you use for the key word *less*?
    (a) Subtraction
    (b) Multiplication
    (c) Division
    (d) Addition

    Answer: _____

13. What is the operation key word(s) in the following sentence?

    Ada received an increase of $48.00.

    Answer: _____

14. What operation will you use for the key words *fraction of*?
    (a) Division
    (b) Subtraction
    (c) Addition
    (d) Multiplication

    Answer: _____

15. What is the operation key word(s) in the following sentence?

    Maya visits her doctor every month.

    Answer: _____

16. What operation will you use for the key word(s) in question 15?
    (a) Division
    (b) Multiplication
    (c) Addition
    (d) Subtraction

    Answer: _____

1.4 Operation Key Words 2 (*)

Example 1:

What operation will you use for the key word *times*?

    (a) Subtraction
    (b) Multiplication
    (c) Addition
    (d) Division

Solution:

The key word *times* is a multiplication key word. So the answer is (b).

Example 2:

What is the operation key word(s) in the following sentence?

    Soham received a raise of $568.43.

Solution:

In this sentence, *raise of* are addition key words.

Write or choose the letter of the answer.

1. What is the operation key word(s) in the following sentence?

    The car can travel 200 kilometers per hour.

                Answer: _____

2. What operation will you use for the key word(s) in question 1?

    (a) Multiplication
    (b) Addition
    (c) Division
    (d) Subtraction

                Answer: _____

3. What is the operation key word(s) in the following sentence?

    Find 50 plus 21.

                Answer: _____

4. What operation will you use for the key word(s) in question 3?

    (a) Division
    (b) Subtraction
    (c) Addition
    (d) Multiplication

                Answer: _____

5. What operation will you use for the key word *remainder*?

    (a) Subtraction
    (b) Division
    (c) Addition
    (d) Multiplication

                Answer: _____

6. What operation will you use for the key words *in all*?

    (a) Addition
    (b) Multiplication
    (c) Subtraction
    (d) Division

                Answer: _____

Write or choose the letter of the answer.

7. What operation will you use for the key words *total of*?
    (a) Addition
    (b) Division
    (c) Subtraction
    (d) Multiplication

    Answer: _____

8. What is the operation key word(s) in the following sentence?

    One-fifth of Maria's salary is donated to charity.

    Answer: _____

9. What operation will you use for the key word(s) in question 8?
    (a) Addition
    (b) Multiplication
    (c) Division
    (d) Subtraction

    Answer: _____

10. What is the operation key word(s) in the following problem?

    Laura had 14 blue pens and 11 black pens. How many pens did she have in total?

    Answer: _____

11. What is the operation key word(s) in the following sentence?

    They needed 3 kilograms of rice tripled for a party.

    Answer: _____

12. What is the operation key word(s) in the following expression?

    $76.00 divided equally among 4 friends

    Answer: _____

13. What operation will you use for the key word(s) in question 12?
    (a) Multiplication
    (b) Subtraction
    (c) Division
    (d) Addition

    Answer: _____

14. What operation will you use for the key words *product of*?
    (a) Subtraction
    (b) Division
    (c) Addition
    (d) Multiplication

    Answer: _____

15. What operation will you use for the key word *twice*?
    (a) Subtraction
    (b) Multiplication
    (c) Addition
    (d) Division

    Answer: _____

16. What is the operation key word(s) in the following sentence?

    Find the product of 4.5 and 1.2.

    Answer: _____

1.5 Write a Math Expression (*)

Example 1:

Rob walked $\dfrac{3}{4}$ of a kilometer from

home to the store and $\dfrac{1}{4}$ of a kilometer

from the store to his friend's house. How far did Rob walk in total?

Which math sentence will you use to find the answer?

  (a) $\dfrac{3}{4} + \dfrac{1}{4}$

  (b) $\dfrac{3}{4} \times \dfrac{1}{4}$

  (c) $\dfrac{3}{4} - \dfrac{1}{4}$

  (d)   All of the above

Solution:

The following information is given:

    Distance walked from home to store =

                     $\dfrac{3}{4}$ kilometers

    Distance walked from store to friend's

    house = $\dfrac{1}{4}$ kilometers

You can find the math sentence as follows:

    (Total distance Rob walked)

      = (Distance walked from home to store) + (Distance walked from store to friend's house)

      = $\dfrac{3}{4} + \dfrac{1}{4}$

So the answer is (a).

Example 2:

One kilogram of mangoes cost $4.00. What is the cost of 6 kilograms of mangoes?

Which math sentence will you use to find the answer?

  (a) $\$4.00 \div 6$

  (b) $\$4.00 \times 6$

  (c) $\$4.00 + 6$

  (d) $\$4.00 - 6$

Solution:

You can find the math sentence as follows:

    Cost of 1 kilogram of mangoes = $4.00

    Cost of 6 kilograms of mangoes

      = (Cost of 1 kilogram of mangoes)

                   × (Total weight)

    = $\$4.00 \times 6$

So the answer is (b).

Write or choose the letter of the answer.

1. What is the operation key word(s) in the following question?

    Find the product of 8.5 and 2.4.

    Answer: _____

2. What is the math sentence for question 1?
    (a) $8.5 \div 2.4$
    (b) 8.5×2.4
    (c) $8.5 + 2.4$
    (d) $8.5 - 2.4$

    Answer: _____

3. What is the operation key word(s) in the following sentence?

    What number do you get if 25.75 is taken away from 41.26?

    Answer: _____

4. A man ran 25 kilometers in 2 hours. Find the distance traveled per hour.

    Which math sentence will you use to find the answer?
    (a) 25×2
    (b) $25 + 2$
    (c) $25 - 2$
    (d) None of the above

    Answer: _____

5. What is the operation key word(s) in the following sentence?

    Juhi has 12 more candies than Daisy.

    Answer: _____

6. The 8 toys cost $3.12. What is the cost of each toy?

    Which math sentence will you use to find the answer?
    (a) $3.12 \div 8$
    (b) 3.12×8
    (c) $3.12 + 8$
    (d) All of the above

    Answer: _____

7. 28 cookies divided equally among 7 girls

    Which math sentence will you use to find the answer?
    (a) $28 - 7$
    (b) $28 + 7$
    (c) $28 \div 7$
    (d) All of the above

    Answer: _____

8. Find the total of 9.6 and 13.8.

    Which math sentence will you use to find the answer?
    (a) $9.6 \div 13.8$
    (b) $9.6 + 13.8$
    (c) $9.6 - 13.8$
    (d) All of the above

    Answer: _____

9. One chocolate costs $5.25. Which math sentence will you use to find the cost of 6 chocolates?
    (a) $5.25 \div 6$
    (b) 5.25×6
    (c) $5.25 + 6$
    (d) All of the above

    Answer: _____

1.6 Write a Math Expression with Multiple Operations (*)

<u>Example 1</u>:

The number 28 is multiplied by the sum of 12 and 7. Which math sentence can you use to find the answer?

(a) $(12 + 7) \times 28$
(b) $(12 - 7) \div 28$
(c) $(12 + 7) - 28$
(d) $(12 \times 7) + 28$

<u>Solution</u>:

You can write the math sentence as follows:

- Sum of 12 and 7 = $(12 + 7)$
- 28 is multiplied by (the sum of 12 and 7)

    = 28 is multiplied by $(12 + 7)$
    = $(12 + 7) \times 28$

So the answer is (a).

<u>Example 2</u>:

What is the math sentence for the following expression?

2.4 is decreased by 25 percent of 20

(a) $(0.25 \times 20) - 2.4$
(b) $(0.25 \times 20) - 2.4$
(c) $(0.25 \div 20) + 2.4$
(d) $(0.25 \times 20) \div 2.4$

<u>Solution</u>:

You can write the math sentence as follows:

- 25% of 20 = (25% times 20)
    = 0.25 times 20
    = (0.25×20)
- 2.4 decreased by (25% of 20)
    = (25% of 20) - 2.4
    = $(0.25 \times 20) - 2.4$

So the answer is (b).

Choose the letter of the answer.

1. 7,000 added to the product of 200 and 15.

   Which math sentence can you use to find the answer?

   (a) $(200 \times 15) + 7,000$
   (b) $(200 \div 15) - 7,000$
   (c) $200 \times (15 - 7,000)$
   (d) $(200 \times 15) - 7,000$

   Answer: _____

2. What is the math sentence for the following expression?

   Multiply 35 by the sum of 124 and 101.

   (a) $(124 + 101) \times 35$
   (b) $(124 \div 101) \times 35$
   (c) $(124 \times 101) - 35$
   (d) $124 \times (101 - 35)$

   Answer: _____

Choose the letter of the answer.

3. 6.7 taken away from the product of 8 and 11.3.

   Which math sentence can you use to find the answer?

   (a) $(8 + 11.3) \times 6.7$

   (b) $(8 + 11.3) - 6.7$

   (c) $(8 \times 11.3) + 6.7$

   (d) $(8 \times 11.3) - 6.7$

   Answer: _____

4. Subtract 22 from the quotient of 315 and 5.

   Which math sentence can you use to find the answer?

   (a) $(315 \times 5) - 22$

   (b) $(315 \div 5) - 22$

   (c) $(315 \times 5) + 22$

   (d) $5 + (315 - 22)$

   Answer: _____

5. What is the math sentence for the following expression?

   $2\dfrac{5}{7}$ taken away from the quotient of 27 and $\dfrac{1}{7}$

   (a) $\left(27 \div \dfrac{1}{7}\right) + 2\dfrac{5}{7}$

   (b) $\left(27 \div \dfrac{1}{7}\right) - 2\dfrac{5}{7}$

   (c) $\left(27 \div \dfrac{1}{7}\right) \times 2\dfrac{5}{7}$

   (d) $\left(27 \div \dfrac{1}{7}\right) \div 2\dfrac{5}{7}$

   Answer: _____

6. What is the math sentence for the following expression?

   12 fewer than 14 tripled

   (a) $(3 \times 14) + 12$

   (b) $(3 \times 14) - 12$

   (c) $(3 \div 14) - 12$

   (d) $(3 + 14) + 12$

   Answer: _____

7. What is the math sentence for the following expression?

   26 more than the sum of 160 and 48

   (a) $(160 \div 48) - 26$

   (b) $(160 \div 48) + 26$

   (c) $(160 \times 48) - 26$

   (d) $(160 + 48) + 26$

   Answer: _____

8. $\dfrac{1}{4}$ of 20 divided by $\dfrac{1}{6}$

   Which math sentence can you use to find the answer?

   (a) $\left(\dfrac{1}{4} + 20\right) - \dfrac{1}{6}$

   (b) $\left(\dfrac{1}{4} \times 20\right) \div \dfrac{1}{6}$

   (c) $\left(\dfrac{1}{4} + 20\right) + \dfrac{1}{6}$

   (d) $\left(\dfrac{1}{4} \times 20\right) \times \dfrac{1}{6}$

   Answer: _____

1.7 Review of Chapter 1 (*)

Write or choose the letter of the answer.

1. What is the operation key word(s) in the following expression?

   Difference between 52 and 41

   Answer: _____

2. What operation will you use for the key word(s) in question 1?
   (a) Multiplication
   (b) Addition
   (c) Division
   (d) Subtraction

   Answer: _____

3. What operation will you use for the key words *total of*?
   (a) Subtraction
   (b) Multiplication
   (c) Addition
   (d) Division

   Answer: _____

4. What is the operation key word(s) in the following sentence?

   Find one-fifth of 2,400.

   Answer: _____

5. What operation will you use for the key word *triple*?
   (a) Division
   (b) Multiplication
   (c) None of the above

   Answer: _____

6. What is the operation key word(s) in the following expression?

   10 fewer than 19

   Answer: _____

7. What operation will you use for the key word *loss*?
   (a) Addition
   (b) Division
   (c) Subtraction
   (d) Multiplication

   Answer: _____

8. What operation will you use for the key word *gain*?
   (a) Subtraction
   (b) Division
   (c) Addition
   (d) Multiplication

   Answer: _____

9. Which math sentence will you use to find 15 tripled?
   (a) $3 \div 15$
   (b) 3×15
   (c) $3 + 15$
   (d) All of the above

   Answer: _____

Write or choose the letter of the answer.

10. Which math sentence will you use to find the answer?

   Sneha had 4 pairs of earrings. She bought 2 more. How many pairs of earrings does Sneha have now?

   (a) $4 + 2$

   (b) $4 - 2$

   (c) 4×2

   (d) All of the above

   Answer: _____

11. The total of $240.00 and $120.00 divided equally among 5 employees.

   Which math sentence will you use for the above expression?

   (a) $(\$240.00 + \$120.00) \times 5$

   (b) $(\$240.00 \div \$120.00) - 5$

   (c) $(\$240.00 + \$120.00) \div 5$

   (d) $(\$240.00 \times \$120.00) + 5$

   Answer: _____

12. What is the math sentence for the following expression?

   Total of 23 and 24 candies are distributed equally among 9 kids

   (a) $(23 + 24) \div 9$

   (b) $(23 \times 24) - 9$

   (c) $(23 \div 24) + 9$

   (d) $(23 + 24) \times 9$

   Answer: _____

13. What is the math sentence for the following expression?

   $\frac{2}{9}$ is added to $\frac{1}{8}$ doubled

   (a) $\left(2 + \frac{1}{8}\right) \times \frac{2}{9}$

   (b) $\left(2 \times \frac{1}{8}\right) - \frac{2}{9}$

   (c) $\left(2 \div \frac{1}{8}\right) \times \frac{2}{9}$

   (d) $\left(2 \times \frac{1}{8}\right) + \frac{2}{9}$

   Answer: _____

14. $\frac{1}{3}$ of 12 multiplied by $\frac{1}{7}$

   Which math sentence can you use for the above expression?

   (a) $\left(\frac{1}{3} + 12\right) - \frac{1}{7}$

   (b) $\left(\frac{1}{3} \times 12\right) \times \frac{1}{7}$

   (c) $\left(\frac{1}{3} \times 12\right) + \frac{1}{7}$

   (d) All of the above

   Answer: _____

15. What is the operation key word(s) in the following expression?

   Sum of 12 and 16

   Answer: _____

2. Basic Problem-Solving Strategies

2.1 One-Step Problems (*)

Example 1:

Vijay bought 7 shirts. He paid $210.00 for the shirts. What was the cost of one shirt?

Solution:

The following information is given:

Number of shirts = 7

Total cost = $210.00

You can find the cost of one shirt by dividing the total cost by the number of shirts.

(cost of one shirt)
     = (total cost)
         ÷ (number of shirts)
     = $210.00 ÷ 7
     = $30.00

So the cost of one shirt was $30.00.

Write or choose the letter of the answer.

1. A house has 7 rooms. Each room has 2 windows. What operation will you use to find the total number of windows?
   - (a) Subtraction
   - (b) Division
   - (c) Multiplication
   - (d) Addition

Answer: _____

Example 2:

What is the operation key word(s) in the following problem?

Daniel has 10 candies. Sarah has 2 fewer candies than Daniel. How many candies does Sarah have?

Solution:

The key word *fewer* is the subtraction key word.

2. Luke wants to complete a project. He works for 7 hours on Saturday and 8 hours on Sunday. What operation will you use to find the total time taken to complete the project?
   - (a) Addition
   - (b) Multiplication
   - (c) Division
   - (d) Subtraction

Answer: _____

Write or choose the letter of the answer.

3. Disha can drink 3 bottles of milk in 1 week. Nancy can drink 4 bottles of milk in the same time. What is the total number of milk bottles required for both of them in 1 week?

Answer: _____ _____
unit

4. Neeraj can write 14 pages in an hour. How many pages can he complete if he writes for 2.5 hours?

Answer: _____ _____
unit

5. If one bag can hold 5 kilograms of wheat, how many bags can be filled with 35 kilograms of wheat?

Answer: _____ _____
unit

6. Lisa took 15 minutes to fill a tank. How much time will she take if she has to fill one-third of the tank?

Answer: _____ _____
unit

7. Arun can drive 72 kilometers in 1 hour. Anisha can drive 67 kilometers in the same time. What is the total distance traveled by both of them in 1 hour?

Answer: _____ _____
unit

8. There are 8 workers assigned to clean an apartment. If they make $176.00 in total, what amount do they each receive?

Answer: _____

9. The cost of a backpack is $11.50. Trevor sold 6 backpacks. What is the total amount paid for the backpacks?

Answer: _____

10. What is the operation key word(s) in the following problem?

    Kevin earned $900.00. Natasha earned $70.00 less than Kevin.

Answer: _____

11. What operation will you use for the key word(s) in question 10?
    (a) Subtraction
    (b) Division
    (c) Multiplication
    (d) Addition

Answer: _____

12. Brendon can pick 72 flowers in an hour. How many flowers can he pick in 20 minutes?

Answer: _____ _____
unit

2.2 Multistep Problems 1 (**)

Example 1:

George had 150 kilograms of rice in his shop. He sold one-tenth of the rice and kept the rest of the rice for a party.

If you want to find the amount of rice kept for the party, what question do you need to answer first?

  (a) How much rice did George sell?
  (b) How much rice was there in his shop?
  (c) How much rice did George keep for the party?
  (d) All of the above

Solution:

You need to find the amount of rice sold before you can find the amount of rice kept for party.

So the answer is (a).

Example 2:

How much rice did George sell in example 1?

Solution:

You can solve the problem as follows:

Amount of rice in shop = 150 kilograms

Amount of rice sold

$$= \frac{1}{10} \text{ of (amount of rice in shop)}$$

$$= \frac{1}{10} \text{ of } 150$$

$$= \frac{1}{10} \times 150 = 15 \text{ kilograms}$$

So George sold 15 kilograms of rice.

Example 3:

Jiten spent $70.00 in total. He spent 0.5 of the money to buy some groceries, $17.00 to buy books, and the rest to buy perfume. How much money did he spend on perfume?

Solution:

The following information is given:

Total money spent = $70.00

Money spent on groceries
  = 0.5 of (total money spent)
  = 0.5 of $70.00

Money spent on books = $17.00

You can use the following steps to find the answer:

Step 1: Find the amount of money spent on groceries.

Money spent on groceries
  = 0.5 of $70.00
  = 0.5 × 70 = $35.00

Step 2: Find the amount of money spent on groceries and books.

Money spent on groceries and books
  = (money spent on groceries)
    + (money spent on books)
  = $35.00 + $17.00 = $52.00

Step 3: Find the money spent on perfume.

Money spent on perfume
  = (total money spent) − (money spent on groceries and books)
  = $70.00 − $52.00
  = $18.00

So Jiten spent $18.00 on perfume.

Write or choose the letter of the answer.

1. Alice had $10.00 to buy fruit. She spent $\frac{3}{5}$ of the money on bananas and the rest on oranges. How much money did Alice spend on oranges?

   Answer: _____

2. Kavya spent $60.00 in total. She spent 0.4 of the money on a backpack, $16.00 on biscuits, and the rest on magazines. How much money did she spend on magazines?

   Answer: _____

3. A shopkeeper had 120 flowers in his shop. There were 44 roses, 41 marigolds, and the rest were lotus flowers. How many lotus flowers were there?

   Answer: ____ ___unit___

4. Mr. Taylor bought some medicine. He paid $9.84 for cough syrup and $12.21 for aspirin. He was left with the same amount of money as he spent at the store. How much did he have at the beginning?

   Answer: _____

5. Akhil had $76.00. He spent half the money on clothes and the rest on writing supplies.

   To find the money spent on writing supplies, what question do you need to answer first?
   (a) How much money did Akhil spend on writing supplies?
   (b) How much money did Akhil have at the beginning?
   (c) How much money did Akhil spend on clothes?
   (d) All of the above

   Answer: _____

6. Bob has to cover 45 kilometers to reach his office. He has already covered $\frac{4}{5}$ of the distance.

   To find the remaining distance, what question do you need to answer first?
   (a) How much distance does Bob have to cover in total?
   (b) How much distance has he already covered?
   (c) How much distance does he want to cover?
   (d) All of the above

   Answer: _____

7. There are 400 students in a school. Out of them, 0.6 of the students are boys and the rest are girls. How many girls are there in the school?

   Answer: ____ _____
                     unit

2.3 Multistep Problems 2 (***)

Example 1:

Lucy had 150 coins from different countries. She donated 0.6 of the coins to a museum.

What question do you need to answer first to find the number of coins left after donating to the museum?

(a) How many coins does Lucy have?

(b) How many coins were left?

(c) How many coins did Lucy donate to the museum?

(d) All of the above

Solution:

You need to find the number of coins donated to the museum before you can find the number of coins left.

So the answer is (c).

Example 2:

How many coins are left after donating to the museum in example 1?

Solution:
You can solve the problem as follows:

Total number of coins = 150

Coins donated to museum

= 0.6 of (total number of coins)

= 0.6 of 150

= 0.6 × 150 = 90 coins

Number of coins left

= total number of coins

– coins donated to museum

= 150 – 90 = 60 coins

So Lucy has 60 coins left after donating to the museum.

Example 3:

The cost of 1 burger is $5.50. Anil bought 8 burgers and gave $50.00 to the cashier. How much money will the cashier return?

Solution:

The following information is given:

Cost of 1 burger = $5.50

Number of burgers = 8

Amount given to the cashier = $50.00

You can find the answer as given below.

Cost of 8 burgers

= (cost of 1 burger)

× (number of burgers)

= $5.50 × 8

= $44.00

Amount returned by the cashier

= (amount given to the cashier)

– (cost of 8 burgers)

= $50.00 – $44.00 = $6.00

So the cashier will return $6.00.

Write or choose the letter of the answer.

1. Carl went to a fair with his friends. He bought 5 toys at a cost of $4.10 per toy. He also bought some snacks worth $4.50. How much money did he spend in total?

   Answer: _____

2. The cost of 1 pizza is $14.75. Bijay delivered 10 pizzas to an apartment. How much money will Bijay return if he received $150.00 for all the pizzas?

   Answer: _____

3. Bill spent 6 hours studying. He studied math for 2 hours, science for 2.5 hours, and spent the rest of the time on English. How much time did he spend on English?

   Answer: ____ _____
                unit

4. The cost of a storybook is $6.25. Arnav bought 4 storybooks from a store. How much money will the cashier return if Arnav gave him $30.00?

   Answer: _____

5. A rice cooker costs $55.50. Mrs. Wilson bought 2 rice cookers. What amount will the cashier return if she gave $120.00 to the cashier?

   Answer: _____

6. A pack of noodles costs $14.20. Ana bought 4 packs of noodles and gave $60.00 to the cashier.

   What question do you need to answer first to find how much money the cashier will return?
   (a) How many packs of noodles did Ana buy?
   (b) How much did one pack of noodles cost?
   (c) What is the cost of 4 packs of noodles?
   (d) All of the above

   Answer: _____

7. A wallet costs $13.25. David bought 3 wallets and gave $40.00 to the cashier.

   What question do you need to answer first to find how much money the cashier will return?
   (a) How much does one wallet cost?
   (b) How many wallets are there in the store?
   (c) What is the cost of 3 wallets?
   (d) All of the above

   Answer: _____

2.4 Multistep Problems 3 (***)

Example 1:

Nitish bought 60 pieces of candy from a shop. He gave two-thirds of the candy to his friends.

What question do you need to answer first to find how many pieces of candy are left?

    (a) How many pieces of candy are left?

    (b) How many pieces of candy did he give to his friends?

    (c) How many pieces of candy did he buy?

    (d) All of the above

Solution:

You need to find how many pieces of candy Nitish gave to his friends before you can find how many candies are left.

So the answer is (b).

Example 2:

How many pieces of candy are left in example 1?

Solution:

You can solve the problem as follows:
Pieces of candy bought from shop = 60
Pieces of candy given to friends

$$= \frac{2}{3} \text{ of (pieces of candy bought)}$$

$$= \frac{2}{3} \text{ of } 60 = \frac{2}{3} \times 60$$

$$= 40 \text{ pieces of candy}$$

Pieces of candy left
    = (pieces of candy bought from shop)
      − (pieces of candy given to friends)
    = 60 − 40 = 20 pieces of candy

So 20 pieces of candy are left.

Example 3:

Mark got $96.00 from his mother. He paid one-fourth of the amount to his friend and $25.00 for a coaching fee. He returned the remaining amount to his mother. How much money did he pay in total?

Solution:

The following information is given:
    Money got from mother = $96.00

$$\text{Money paid to friend} = \frac{1}{4} \text{ of (money got)}$$

$$= \frac{1}{4} \text{ of } \$96.00$$

    Money paid for coaching fee = $25.00

You can find the answer as shown below.

$$\text{Money paid to friend} = \frac{1}{4} \text{ of } \$96.00$$

$$= \frac{1}{4} \times \$96.00$$

$$= \$24.00$$

Money paid in total
    = (money paid to friend)
       + (money paid for coaching fee)
    = $25.00 + $24.00
    = $49.00

So Mark paid $49.00 in total.

Write or choose the letter of the answer.

1. One bag of beans weighs 4.5 pounds. A shopkeeper bought 8 bags of beans and sold 32 pounds. How much did the shopkeeper buy?

   Answer: ____ _____

unit

2. What amount of beans is left after what the shopkeeper sold in question 1?

   Answer: ____ _____

unit

3. Nikhil bought 6 jars of honey worth $36.00 per jar and sold them for $225.00. How much profit did he make?

   Answer: _____

4. The cost of 1 cake was $17.20. Arun bought 5 cakes and gave $90.00 to the cashier.

   What question do you need to answer first to find out how much money the cashier returned to Arun?

   (a) How many cakes did Arun buy?
   (b) How much did he give to the cashier?
   (c) What was the cost of 5 cakes?
   (d) All of the above

   Answer: _____

5. Mr. Watson's monthly salary is $1,800.00. He spends one-fourth of his salary on clothes, pays $720.00 for his house loan, and spends the rest of the balance on household expenses. How much money does he pay for household expenses?

   Answer: _____

6. Mrs. Carter bought 8 boxes of snacks, each weighing 1.5 pounds. She repacked the snacks equally into 24 small boxes. What is the weight of snacks Mrs. Carter bought?

   Answer: ____ _____

unit

7. What is the weight of each box Mrs. Carter repacked in question 6?

   Answer: ____ _____

unit

8. One jar of flour weighs 1.5 pounds. Mr. Hawkins had 2.5 pounds of flour and bought 10 new jars of flour. If he used 16 pounds of flour, how much flour is left?

   Answer: ____ _____

   unit

2.5 Work Backward 1 (**)

Example 1:

Mrs. Jones bought 5 bowls for $2.50 each, 4 health drinks for $17.50 each, and 2 pillows for $24.50 each. If the cashier returned $3.50, how much money did Mrs. Jones give to the cashier?

Solution:

The following information is given:

Cost of 1 bowl = $2.50

Cost of 1 health drink = $17.50

Cost of 1 pillow = $24.50

Money returned by cashier = $3.50

You can use the following steps to find the answer.

- Find the cost of 5 bowls, 4 health drinks, and 2 pillows.

(cost of 5 bowls) = $2.50 × 5 = $12.50

(cost of 4 health drinks)

= $17.50 × 4 = $70.00

(cost of 2 pillows)

= $24.50 × 2 = $49.00

- Find the cost of all supplies.

(cost of all supplies)

= $12.50 + $70.00 + $49.00 = $131.50

- Find the amount given to the cashier.

(amount given to the cashier)

= $131.50 + $3.50 = $135.00

So Mrs. Jones gave $135.00 to the cashier.

Example 2:

Michelle and her friends spent $60.00 in a store. They bought 5 shirts that cost the same amount and a box of cookies for $5.00. What was the cost of each shirt?

Solution:

The following information is given:

Total amount spent = $60.00

Cost of a box of cookies = $5.00

Number of shirts = 5

You can solve this problem by working backward.

Step 1: Find the cost of 5 shirts.

(cost of 5 shirts)

= (total amount spent)

− (cost of cookies)

= $60.00 − $5.00 = $55.00

Step 2: Find the cost of each shirt.

(cost of each shirt)

= (cost of 5 shirts)

÷ (number of shirts)

= $55.00 ÷ 5 = $11.00

So the cost of each shirt was $11.00.

Write the answer.

1. A girl was 60 inches tall in the year 2013. She had grown a total of 11 inches from 2006 to 2013. She had grown 8 inches from 2001 to 2006. How tall was the girl in the year 2001?

    Answer: ____ _____
    unit

2. Maria bought 5 ribbons for $0.20 each, 4 thermoses for $14.00 each, and 3 soaps for $27.00. If the cashier returned $6.00, how much money did Maria give to the cashier?

    Answer: _____

3. Disha spent $49.00 at a store. She bought 6 skirts that cost the same amount and a world map for $7.00. What was the cost of each skirt?

    Answer: _____

4. A plant was 10 inches tall on Saturday. It had grown 3 inches from Thursday to Saturday. It had grown 2 inches from Tuesday to Thursday. How tall was the plant on Tuesday?

    Answer: ____ _____
    unit

5. Kapil went to buy gym equipment. He gave a total of $76.35 and got back $3.65 in change from the cashier. How much money did Kapil give to the cashier?

    Answer: : _____

6. A group of custodians cleaned 150 rooms in total. They cleaned an equal number of rooms in any 5 days of the week and cleaned 25 rooms on the weekend. How many rooms did they clean each weekday?

    Answer: ____ _____
    unit

7. Mrs. Murphy went to buy kitchen appliances. She paid a total of $42.40 for the appliances and got back $2.60 in change from the cashier. How much money did Mrs. Murphy give to the cashier?

    Answer: _____

8. Angela and her friends spent $132.00 at a shopping mall. They bought 4 pairs of jeans that cost the same amount and some snacks for $8.00. What was the cost of each pair of jeans?

    Answer: _____

2.6 Use Variables to Solve Problems 1 (**)

Example 1:

The sum of two numbers is 23. The difference between the two numbers is 1. Use the variable N to represent the first number.

Which equation represents the given problem?

   (a) $2N - 1 = 23$
   (b) $N + 1 = 23$
   (c) $2N + 1 = 23$
   (d) $N = 1 + 23$

Solution:

Use the variable N to represent the first number.

You can write the equation as given below.

First number = N
Second number = (first number) − 1
             = $N - 1$
(first number) + (second number) = 23

$N + N - 1 = 23$

$2N - 1 = 23$

So the answer is (a).

Example 2:

Find the first number in example 1.

Solution:

Solve the equation to find the first number (N).

$2N - 1 = 23$ ← from example 1
$2N - 1 + 1 = 23 + 1$ ← add 1
$2N = 24$
$(2N ÷ 2) = (24 ÷ 2)$ ← divide by 2
$N = 12$

The first number is 12.

Example 3:

Find the second number in example 1.

Solution:

Find the second number ($N - 1$).

$N = 12$ ← from example 2
$N - 1 = 12 - 1 = 11$

The second number is 11.

Write or choose the letter of the answer.

1. The sum of Arushi and her sister's ages is 27. Her sister is 3 years older than Arushi. Use the variable G to represent Arushi's age. Which equation represents this problem?

   (a) $2G - 3 = 27$
   (b) $G + 3 = 27$
   (c) $G - 3 = 27$
   (d) $2G + 3 = 27$

Answer: _____

2. How old is Arushi in the problem in question 1?

Answer: ____ _____
               unit

Write or choose the letter of the answer.

3. Mr. Lewis has 80 pets on his farm. The number of rabbits is 20 more than the number of sheep. Use the variable Y to represent the number of sheep. Which equation represents the given problem?

    (a) $Y - 20 = 80$
    (b) $Y + 20 = 80$
    (c) $2Y - 20 = 80$
    (d) $2Y + 20 = 80$

Answer: _____

4. How many sheep are there on the farm for the problem in question 3?

Answer: ____ _____
                unit

5. Find the number of rabbits on the farm for the problem in question 3.

Answer: ____ _____
                unit

6. There are 140 computers in a lab. The number of laptops is 36 less than the number of desktops. Use the variable X to represent the number of desktops. Which equation represents this problem?

    (a) $2X - 36 = 140$
    (b) $X + 36 = 140$
    (c) $2X + 36 = 140$
    (d) $X - 140 = 36$

Answer: _____

7. Alka has a total of 42 candies and cookies. The number of candies is 8 more than the number of cookies. Use the variable M to represent the number of cookies. Which equation represents this problem?

    (a) $2M - 8 = 42$
    (b) $2M + 8 = 42$
    (c) $M + 8 = 42$
    (d) $M = 8 + 42$

Answer: _____

8. The sum of two numbers is 55. The difference between the two numbers is 7. Use the variable P to represent the first number.

Which equation represents the given problem?

    (a) $2P - 7 = 55$
    (b) $P + 7 = 55$
    (c) $2P + 7 = 55$
    (d) $P - 7 = 55$

Answer: _____

9. Find the first number in the problem in question 8.

Answer: _____

10. Find the second number in the problem in question 8.

Answer: _____

2.7 Use Variables to Solve Problems 2 (**)

Example 1:

The sum of Jacob's and his mom's ages is 59. His mom is 29 years older than Jacob. Use the variable N to represent Jacob's age. Which equation represents the given problem?

(a) $2N + 29 = 59$
(b) $N + 29 = 59$
(c) $2N - 29 = 59$
(d) $N - 29 = 59$

Solution:

Use the variable N to represent Jacob's age.

You can write the equation as given below.

Jacob's age = N
Jacob's mom's age
        = (Jacob's age) + 29
        = $N + 29$
(Jacob's age) + (Jacob's mom's age) = 59
$N + N + 29 = 59$
$2N + 29 = 59$

So the answer is (a).

Example 2:

How old is Jacob in example 1?

Solution:

Find Jacob's age (N).

$2N + 29 = 59$ ← from example 1
$2N + 29 - 29 = 59 - 29$ ← subtract 29
$2N = 30$
$(2N \div 2) = (30 \div 2)$ ← divide by 2
$N = 15$ years

Jacob is 15 years old.

Example 3:

How old is Jacob's mom in example 1?

Solution:

Find the age of Jacob's mom ($N + 29$).

$N = 15$ ← from example 2
$N + 29 = 15 + 29 = 44$ years

Jacob's mom is 44 years old.

Write or choose the letter of the answer.

1. The sum of two numbers is 52. The difference between two numbers is 10. Use the variable X to represent the first number, and find the equation that represents the problem.

(a) $X - 10 = 52$
(b) $X + 10 = 52$
(c) $2X + 10 = 52$
(d) $2X - 10 = 52$

Answer: _____

2. What is the first number in the problem in question 1?

Answer: _____

3. What is the second number in the problem in question 1?

Answer: _____

Write or choose the letter of the answer.

4. A hostel has 105 boys. The number of boys on the first floor is 15 more than the number of boys on the second floor. Use the variable R to represent the number of boys on the second floor. Which equation represents the given problem?

    (a) $120R = 105$
    (b) $R + 15 = 105$
    (c) $2R = 105$
    (d) $2R + 15 = 105$

Answer: _____

5. Find the number of boys on the second floor in the problem in question 4.

Answer: _____ _____
                     unit

6. Anil stored rice and wheat in two storerooms. There were a total of 150 bags of both crops. The number of wheat bags was 10 fewer than the number of rice bags. Use the variable M for the number of rice bags. Which equation represents the given problem?

    (a) $2M = 150$
    (b) $2M - 10 = 150$
    (c) $M - 10 = 150$
    (d) $2M = 150$

Answer: _____

7. Find the number of wheat bags in the problem in question 6.

Answer: _____ _____
                     unit

8. The sum of two numbers is 19. The difference between two numbers is 3. Use the variable K to represent the first number, and find the equation that represents the problem

    (a) $2K = 19$
    (b) $2K + 3 = 19$
    (c) $2K = 22$
    (d) $K = 22$

Answer: _____

9. The sum of Manoj and his dad's ages is 43. Manoj's dad is 31 years older than Manoj. Use the variable S to represent Manoj's age. Which equation represents the given problem?

    (a) $S + 43 = 31$
    (b) $S - 43 = 31$
    (c) $2S + 31 = 43$
    (d) $2S - 31 = 43$

Answer: _____

2.8 Review of Chapter 2 (**)

Write or choose the letter of the answer.

1. Bimal got $35.00 from his uncle, and he has to spend three-fifths of the money on study materials. How much will he spend on study materials?

   Answer: _____

2. Jay has 240 pairs of shoes in his shop. The number of casual shoes is 24 more than the number of formal shoes. Use the variable N to represent the number of formal shoes. Which equation represents the given problem?

   (a) $N - 24 = 240$
   (b) $2N + 240 = 24$
   (c) $2N + 24 = 240$
   (d) $N + 24 = 240$

   Answer: _____

3. Find the number of casual shoes in the problem in question 2.

   Answer: ____ _____
                  unit

4. Gary spent $65.00 at a shopping mall. He bought 2 hats that costs the same amount and shoes for $38.00. What was the cost of each hat?

   Answer: _____

5. What is the operation key word(s) in the following sentence?

   Tiffany received an increase of 9 percent in her monthly salary.

   Answer: _____

6. The sum of two numbers is 55. The difference between the two numbers is 19. Use the variable N to represent the first number.

   Which equation represents the given problem?

   (a) $2N = 36$
   (b) $N + 19 = 55$
   (c) $2N - 19 = 55$
   (d) $N = 74$

   Answer: _____

7. The sum of Brian and his dad's ages is 39. His dad is 29 years older than Brian. Use the variable K to represent Brian's age. Which equation represents this problem?

   (a) $2K + 29 = 39$
   (b) $2K - 39 = 29$
   (c) $K + 39 = 29$
   (d) $K - 29 = 39$

   Answer: _____

Write or choose the letter of the answer.

8. A suitcase costs $42.40. Aditya bought 2 suitcases and gave $90.00 to the cashier.

   Which question do you need to answer first to find how much money the cashier will return?
   (a) How much did one suitcase cost?
   (b) How many suitcases are there in the store?
   (c) What was the cost of 2 suitcases?
   (d) All of the above

   Answer: _____

9. Nancy bought 3 pens for $0.35 each, 6 chocolates for $0.80 each, and 3 notebooks for $3.30 from a store. If the cashier returned $0.85, how much money did Nancy give to the cashier?

   Answer: _____

10. The cost of a juice bottle is $18.00. Neil wants to buy 7 of them. What operation will you use to find the total cost of all the juice bottles?
    (a) Addition
    (b) Division
    (c) Subtraction
    (d) Multiplication

    Answer: _____

11. Mrs. Lopez went to buy some food items. She paid a total of $27.70 for the food and got back $2.30 in change from the cashier. How much money did Mrs. Lopez give to the cashier?

    Answer: _____

12. The sum of Kunal's and Pamela's ages is 60. Pamela is 6 years older than Kunal. Use the variable S to represent Kunal's age. Which equation represents the given problem?
    (a) $2S + 6 = 60$
    (b) $2S = 66$
    (c) $S - 6 = 60$
    (d) $S = 66$

    Answer: _____

13. How old is Pamela in the problem in question 12?

    Answer: _____ _____

    unit

14. How old is Kunal in the problem in question 12?

    Answer: _____ _____

    unit

3. Unitary Method

3.1 Unitary Method—Direct Proportion (*)

Example 1:

If 3 rice bags can hold 15 pounds of rice, how much rice can 5 rice bags hold?

Solution:

This problem can be solved using the following steps.

Step 1: Find the amount of rice in 1 bag.

Amount of rice in 3 bags = 15 pounds
Amount of rice in 1 bag = 15 ÷ 3
              = 5 pounds

Step 2: Find the amount of rice in 5 bags.

Amount of rice in 1 bag = 5 pounds
Amount of rice in 5 bags = 5 × 5
              = 25 pounds

So 5 rice bags can hold 25 pounds of rice.

Note:

Fewer rice bags will hold less rice.

More rice bags will hold more rice.

Example 2:

If 4 girls can make 12 cards, how many girls can make 48 cards?

Solution:

This problem can be solved using the following steps.

Step 1: Find the number of girls to make 1 card.

Number of girls to make 12 cards = 4 girls
Number of girls to make 1 card

$$= 4 \div 12$$

$$= \frac{4}{12} = \frac{1}{3} \text{ girls}$$

Step 2: Find the number of girls to make 48 cards.

Number of girls to make 1 card = $\frac{1}{3}$ girls

Number of girls to make 48 cards

$$= \frac{1}{3} \times 48$$

$$= \frac{48}{3} = 16 \text{ girls}$$

So 16 girls can make 48 cards.

Note:

More girls can make more cards.

Fewer girls can make fewer cards.

Write the answer.

1. If 2 jars can hold 8 liters of honey, how much honey can 7 jars hold?

    Answer: _____ _____
    unit

2. If 3 people can dig 1 hole in an hour, how many people will it take to dig 8 holes in an hour?

    Answer: _____ _____
    unit

3. If 5 cows can give 20 liters of milk, how many cows will it take to give 80 liters of milk?

    Answer: _____ _____
    unit

4. If 6 managers can select 8 executives, how many executives can 9 managers select?

    Answer: _____ _____
    unit

5. If 4 workers plant 1 garden in a day, how many workers will it take to plant 3 gardens in a day?

    Answer: _____ _____
    unit

6. If 7 students can write 105 pages, how many students will it take to write 315 pages?

    Answer: _____ _____
    unit

7. If 3 carpenters can make 9 tables, how many tables can 5 carpenters make?

    Answer: _____ _____
    unit

8. If 6 workers can repair a 1-kilometer-long road in a week, how many workers will it take to repair a 5-kilometer-long road in a week?

    Answer: _____ _____
    unit

9. If 10 children can eat 50 pieces of candy, how many children will it take to eat 100 pieces of candy?

    Answer: _____ _____
    unit

10. If 4 baskets can hold 24 kilograms of fruit, how much fruit can 6 baskets hold?

    Answer: _____ _____
    unit

3.2 Unitary Method—Inverse Proportion (*)

<u>Example 1</u>:

It takes 8 people 3 months to build a cottage. How long will 6 people take to build the cottage?

<u>Solution</u>:

This problem can be solved using the following steps.

<u>Step 1</u>: Find the time taken by 1 person to build the cottage.

Time taken by 8 people to build the cottage

= 3 months

Time taken by 1 person to build the cottage

= 3 × 8

= 24 months

<u>Step 2</u>: Find the time taken by 6 people to build the cottage.

Time taken by 1 person to build the cottage

= 24 months

Time taken by 6 people to build the cottage

= 24 ÷ 6

= 4 months

So 6 people will take 4 months to build the cottage.

<u>Note</u>:

More workers will build a cottage in less time.

Fewer workers will build a cottage in more time.

<u>Example 2</u>:

It takes 6 friends 4 hours to wrap some gifts. There were 3 friends who could not come on one day. How long did the other friends take to wrap the gifts?

<u>Solution</u>:

This problem can be solved using the following steps.

<u>Step 1</u>: Find the time taken by 1 friend to wrap the gifts.

Time taken by 6 friends to wrap the gifts

= 4 hours

Time taken by 1 friend to wrap the gifts

= 4 × 6

= 24 hours

<u>Step 2</u>: Find the time taken by 3 friends to wrap the gifts.

Time taken by 1 friend to wrap the gifts

= 24 hours

Time taken by 3 friends to wrap the gifts

= 24 ÷ 3

= 8 hours

So the other friends took 8 hours to wrap the gifts.

<u>Note</u>:

More friends will wrap the gifts in less time.

Fewer friends will wrap the gifts in more time.

Write the answer.

1. It takes 3 pipes 20 minutes to fill a drum. How long will it take 1 pipe to fill the drum?

    Answer: _____ _____
    <u>unit</u>

2. It takes 9 members 4 hours to complete a survey. How long will it take 6 members to complete the survey?

    Answer: _____ _____
    <u>unit</u>

3. It takes 4 vacuum cleaners 3 hours to clean some labs. If 1 vacuum cleaner did not work, how long would the other vacuum cleaners take to clean the labs?

    Answer: _____ _____
    <u>unit</u>

4. It takes 5 students 2 days to arrange books. How many days will 1 student take to arrange the books?

    Answer: _____ _____
    <u>unit</u>

5. It takes 5 taps 6 hours to fill a family pool. If 2 taps got damaged, how long will it take for the other taps to fill the pool?

    Answer: _____ _____
    <u>unit</u>

6. It takes 6 doctors 4 hours to do an operation. How long will it take 4 doctors to finish the operation?

    Answer: _____ _____
    <u>unit</u>

7. It takes 3 students 4 days to prepare a school presentation. How many days will 1 student take to prepare the presentation?

    Answer: _____ _____
    <u>unit</u>

8. It takes 9 plumbers 4 hours to set up some tanks. One day, 3 plumbers went on leave. How long did the other plumbers take to set up tanks that day?

    Answer: _____ _____
    <u>unit</u>

9. It takes 6 architects 7 days to design a model of a plant. How many days will it take 1 architect to design the model of the plant?

    Answer: _____ _____
    <u>unit</u>

10. It takes 10 scientists 4 months to discover a theory. How long will it take 8 scientists to discover the theory?

    Answer: _____ _____
    <u>unit</u>

3.3 Unitary Method—Time Problems (**)

Example 1:

It takes 1 worker 6 days to paint a building. How many days will 3 workers take to paint the same building ?

Solution:

You can solve this problem as follows.

Number of days taken by 1 worker to paint a building = 6   ← given

Number of days taken by 3 workers to paint the building = 6 ÷ 3
            = 2 days

So 3 workers will take 2 days to paint the building.

Example 2:

Molly and her sister take 5 hours to make some toys. How long will it take Molly if she wants to make those toys alone?

Solution:

You can solve this problem as follows.

Time taken by 2 people to make some toys
            = 5 hours
Time taken by 1 person to make the toys
            = 5 × 2
            = 10 hours

So it will take 10 hours if Molly wants to make those toys without her sister.

Example 3:

Jack takes 36 minutes to write 18 pages. How long will he take to write 30 pages?

Solution:

This problem can be solved using the following steps.

Step 1: Find the time taken to write 1 page.

Time taken to write 18 pages = 36 minutes
Time taken to write 1 page = 36 ÷ 18
            = 2 minutes

Step 2: Find the time taken to write 30 pages.

Time taken to write 1 page = 2 minutes
Time taken to write 30 pages = 2 × 30
            = 60 minutes

So Jack will take 60 minutes to write 30 pages.

Note:

Writing more pages will take more time.

Writing fewer pages will take less time.

Write the answer.

1. Bijay takes 4 hours to teach 3 subjects. How long will he take to teach 6 subjects?

    Answer: _____ _____
    _{unit}

2. It takes 1 worker 6 days to make a brick wall. How many days will 2 workers take to make the brick wall?

    Answer: _____ _____
    unit

3. Nil takes 30 minutes to mow a 25-square-meter lawn. How long will he take to mow a 75-square-meter lawn?

    Answer: _____ _____
    unit

4. Nancy and 2 of her brothers take 2 hours to fill a tank using buckets. How long will it take Nancy if she wants to fill the tank alone ?

    Answer: _____ _____
    unit

5. Anil takes 40 minutes to run 6 kilometers. How long will he take to run 15 kilometers?

    Answer: _____ _____
    unit

6. It takes 1 person 6 hours to decorate a hall. How many hours will 4 people take to decorate the hall?

    Answer: _____ _____
    unit

7. Sofia and 3 friends take 3 hours to write a manuscript. How long will it take Sofia if she wants to write the manuscript alone?

    Answer: _____ _____
    unit

8. A machine takes 20 minutes to fill 28 bottles. How long will it take to fill 70 bottles?

    Answer: _____ _____
    unit

9. Peter and 5 friends take 30 days to complete a project. How long will it take Peter if he wants to complete the project alone?

    Answer: _____ _____
    unit

10. It takes 1 child 35 minutes to eat a box of cookies. How many minutes will it take 5 children to eat the cookies?

    Answer: _____ _____
    unit

3.4 Unitary Method—Work Problems (**)

Example 1:

Martin can design a template in 6 hours. What fraction of the template can he design in 1 hour?

Solution:

Consider the whole template as 1 unit.

Number of templates designed in 6 hours
$$= 1$$

Number of templates designed in 1 hour

$$= (1 \div 6) \text{ units} \leftarrow \text{divide by 6}$$

$$= \frac{1}{6} \text{ units}$$

$$= \frac{1}{6} \text{ of the whole template}$$

So Martin can design $\left(\frac{1}{6}\right)$ of the whole template in 1 hour.

Example 2:

A train travels 208 miles in 4 hours. How many miles will it travel in 7 hours?

Solution:

This problem can be solved using the following steps.

Step 1: Find the distance traveled in 1 hour.

Distance traveled in 4 hours = 208 miles

Distance traveled in 1 hour = 208 ÷ 4
$$= 52 \text{ miles}$$

Step 2: Find the distance traveled in 7 hours.

Distance traveled in 1 hour = 52 miles

Distance traveled in 7 hours = 52 × 7
$$= 364 \text{ miles}$$

So the train will travel 364 miles in 7 hours.

Write the answer.

1. Rohit can run 42 miles in 3 days. How many miles can he run in 5 days?

Answer: ____ _____
unit

2. Jessica can read a novel in 7 days. What fraction of the novel can she read in 1 day?

Answer: ____ _____
unit

3. A pump can fill 2 tanks in 40 minutes. How many tanks will it fill in 120 minutes?

Answer: ____ _____
unit

4. Nikhil can finish a task in 3 hours. What fraction of the task can he finish in 1 hour?

Answer: ____ _____
unit

Write the answer.

5. Caroline can sew 15 dresses in 5 days. How many dresses can she sew in 9 days?

    Answer: _____ _____
                          unit

6. Grace can sweep a floor in 22 minutes. What fraction of the floor can she sweep in 1 minute?

    Answer: _____ _____
                          unit

7. Neeraj types 72 words in 4 minutes. How many words will he type in 10 minutes?

    Answer: _____ _____
                          unit

8. A car covers 800 meters in 50 seconds. How much distance will it cover in 30 seconds?

    Answer: _____ _____
                          unit

9. Kavya can water the plants in a garden in 5 hours. What fraction of the garden can she water in 1 hour?

    Answer: _____ _____
                          unit

10. Mrs. Agrawal can prepare a special dish in 45 minutes. What fraction of the dish can she prepare in 1 minute?

    Answer: _____ _____
                          unit

11. Karan solves 6 questions in 15 minutes. How many questions will he solve in 75 minutes?

    Answer: _____ _____
                          unit

12. Adriana can make a painting in 3 hours. What fraction of the painting can she make in 1 hour?

    Answer: _____ _____
                          unit

13. Brian makes 56 toy boats in 7 days. How many toy boats will he make in 12 days?

    Answer: _____ _____
                          unit

14. Mr. Fleming saves $840.00 from his monthly salary in 4 months. How much money will he save in 6 months?

    Answer: _____ _____
                          unit

3.5 Review of Chapter 3 (**)

Write the answer.

1. It takes 2 pumps 40 minutes to empty a tank. How many minutes will it take 1 pump to empty the tank?

    Answer: ____ _____
    unit

2. It takes 3 robots 6 hours to assemble a car. How long will 6 robots take to assemble the car?

    Answer: ____ _____
    unit

3. It takes 1 author 6 days to write a manuscript. How many days will 2 authors take to write the manuscript?

    Answer: ____ _____
    unit

4. Anuj takes 4 hours to write 48 pages. How long will he take to write 36 pages?

    Answer: ____ _____
    unit

5. Sophia and 3 friends take 5 hours to weed a garden. How long will it take Sophia if she wants to weed the garden alone?

    Answer: ____ _____
    unit

6. If 3 cans can hold 27 liters of milk, how much milk can 5 cans hold?

    Answer: ____ _____
    unit

7. If 6 plumbers can set up tanks in 1 building in a day, how many plumbers will set up tanks in 3 buildings in a day?

    Answer: ____ _____
    unit

8. Jasmin went on stage to sing 3 songs in 15 minutes. She forgot 1 song and could not sing. How long did she take to sing the other songs?

    Answer: ____ _____
    unit

9. Andrew writes 2 essays in 40 minutes. How many essays will he write in 100 minutes?

    Answer: ____ _____
    unit

10. If 7 women can make 21 baskets, how many women can make 84 baskets?

    Answer: ____ _____
    unit

Write the answer.

11. Kevin can complete an assignment in 6 hours. What fraction of the assignment can he finish in 1 hour?

Answer: ____ _____
unit

12. Amit and his brother take 4 hours to calculate some data. How long will it take Amit if he wants to calculate the data alone?

Answer: ____ _____
unit

13. It takes 5 friends 3 hours to arrange chairs for a party. However, 2 friends could not come to help. How long will it take for the other friends to arrange the chairs?

Answer: ____ _____
unit

14. If 2 boxes can hold 42 pounds of clothes, what amount of clothes can 7 boxes hold?

Answer: ____ _____
unit

15. If 4 farmers can weed 1 field in a day, how many farmers will it take to weed 4 fields in a day?

Answer: ____ _____
unit

16. It takes 1 kid 18 minutes to eat a pizza. How long will 3 kids take to eat the pizza?

Answer: ____ _____
unit

17. Alex takes 6 hours to design 36 crafts. How long will he take to design 24 crafts?

Answer: ____ _____
unit

18. Stella can compose a song in 35 minutes. What fraction of the song can she compose in 1 minute?

Answer: ____ _____
unit

19. Mr. Wilson can shoot 6 videos in 24 days. How many videos can he shoot in 32 days?

Answer: ____ _____
unit

20. Mrs. Waugh uses a bag of sugar in 7 days. What fraction of the sugar will she use in 1 day?

Answer: ____ _____

unit

4. Number Problems

4.1 Place Value Concepts (**)

Example 1:

What is the sum of the values of 8, 3, and 0 in 82,320?

Solution:

In the number 82,320

Value of 8 = 80,000
Value of 3 = 300
Value of 0 = 0

Sum of the values of 8, 3, and 0
= 80,000 + 300 + 0
= 80,300

So the sum of the values of 8, 3, and 0 is 80,300.

Example 2:

What is the sum of the place values of 7 and 4 in 7,479?

Solution:

In the number 7,479

Place value of 7 = 1,000
Place value of 4 = 100

Sum of the place values of 7 and 4
= 1,000 + 100
= 1,100

So the sum of the place values of 7 and 4 is 1,100.

Example 3:

What is the largest 4-digit number that is divisible by 8?

Solution:

First, write the largest 4-digit number, and see if it is divisible by 8. If it is, you have the answer. If not, you can use the concept of the remainder to find the largest number that is divisible by 8.

• The largest 4-digit number is 9,999.

• 9,999 ÷ 8 = 1,249 R 7. The remainder is 7. So this is not divisible by 8.

• If you subtract the remainder, 7, from 9,999, the difference will be divisible by 8.

• 9,999 − 7 = 9,992. This is divisible by 8.

• So the largest 4-digit number divisible by 8 is 9,992.

Example 4:

What is the largest 3-digit number that is less than 517?

Solution:

The largest number less than 517 is the number that is 1 less than 517.

So the answer is 517 − 1 = 516.

Write the answer.

1. What is the sum of the values of 7, 3, and 1 in 37,831?

   Answer: _____

2. What is the largest 5-digit number that is less than 63,720?

   Answer: _____

3. What is the largest 6-digit number that is less than 471,393?

   Answer: _____

4. What is the sum of the place values of 5, 3, and 0 in 57,630?

   Answer: _____

5. What is the largest 4-digit number that is divisible by 10?

   Answer: _____

6. What is the sum of the place values of 6 and 3 in 6,935?

   Answer: _____

7. What is the largest 4-digit number that is less than 2,795?

   Answer: _____

8. What is the sum of the place values of 2, 6, and 3 in 27,563?

   Answer: _____

9. What is the largest 5-digit number that is divisible by 2?

   Answer: _____

10. What is the sum of the values of 6, 8, and 2 in 46,820?

    Answer: _____

11. What is the largest 6-digit number that is divisible by 6?

    Answer: _____

12. What is the largest 3-digit number that is less than 121?

    Answer: _____

13. What is the largest 3-digit number that is divisible by 8?

    Answer: _____

14. What is the sum of the place values of 8 and 9 in 8,393?

    Answer: _____

4.2 Decimal Place Value Concepts (**)

Example 1:

What is the smallest number with 2 decimal places that is greater than 9?

Solution:

You can find the answer using a decimal place value table.

ones	decimal point	tenths	hundredths

- The number has two decimal places. So place the decimal point in the third box from the right.

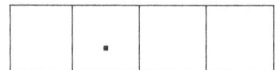

- The number is the smallest number that is greater than 9. This means you have 9 before the decimal point. Any other number before the decimal point will not be the smallest. So write 9 before the decimal point.

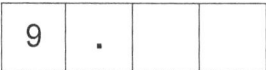

- For the number to be the smallest, all digits after the decimal point have to be 0 except the hundredths digit, which has to be 1. If you write 0 in the hundredths place, the number will not have two decimal places. Try other numbers for the digits, and check if the number is the smallest.

9	.	0	1

So the decimal number is 9.01.

Example 2:

A number has 2 decimal places. It is greater than 6 and less than 7. The tenths-place digit is 3 more than the hundredths-place digit. If the hundredths-place digit is 2, what is the decimal number?

Solution:

You can find the answer using a decimal place value table.

ones	decimal point	tenths	hundredths

- The number has two decimal places. So place the decimal point in the third box from the right.

- The hundredths-place digit is 2. So write 2 in the hundredths place.

	.		2

- The tenths-place digit is 3 more than the hundredths-place digit. So the tenths-place digit is 2 + 3 = 5. Write 5 in the tenths place.

	.	5	2

- The number is greater than 6 and less than 7. That means the ones-place digit must be 6. So write 6 in the ones place.

6	.	5	2

So the decimal number is 6.52.

Write the answer.

1. I am the largest possible decimal number that is less than 14. Each of my digits is one-third of the digit to its right. What number am I?

   Answer: _____

2. What is the difference between the place values of 3 and 4 in 637.24?

   Answer: _____

3. A number has 2 decimal places. It is greater than 11 and less than 12. The tenths-place digit is the same as the hundredths-place digit. If the hundredths-place digit is the largest single digit odd number, what is the decimal number?

   Answer: _____

4. What is the smallest number with three decimal places that is greater than 42?

   Answer: _____

5. What is the difference between the place values of 6 and 2 in 64.203?

   Answer: _____

6. A number has 2 decimal places. It is greater than 32 and less than 33. The tenths-place digit is 2 less than the hundredths-place digit. If the hundredths-place digit is 7, what is the decimal number?

   Answer: _____

7. A number has 3 decimal places. It is greater than 7 and less than 8. The hundredths-place digit is 6 more than the thousandths-place digit, and the tenths-place digit is the same as the hundredths-place digit. If the thousandths-place digit is 3, what is the decimal number?

   Answer: _____

8. What is the smallest number with two decimal places that is greater than 310?

   Answer: _____

4.3 GCF and LCM Basics - 1 (**)

Example 1:

Find the GCF of the following numbers:

| 45, 60, 75

Example 2:

What is the smallest number that must be added to 200 so that the sum is divisible by 10, 14, and 15?

Solution:

The greatest common factor (GCF) can be found using the following steps.

- Divide all the numbers by a common factor.

- Continue dividing by common factors until there is no factor of all the numbers left.

- Multiply only the common factors to find the GCF.

```
5 | 45, 60, 75   ← Divide by 5
3 |  9, 12, 15   ← Divide by 3
      3, 4, 5    ← No more common
                   factors
```

Multiply the common factors 5 and 3 to find the GCF.

GCF of 45, 60, and 75 = 5 × 3 = 15

Note that the GCF of 45, 60, and 75 is the largest number that divides these numbers.

Solution:

First, find the least common multiple (LCM) of 10, 14, and 15. The LCM is the smallest number that is divisible by these numbers.

Then find the number that needs to be added to 200 so the sum is the number found in the first step.

Step 1: Find the LCM of 10, 14, and 15.

- Divide two or more numbers by a common factor.
- Continue dividing until the only common factor is 1.

```
2 | 10, 14, 15   ← Divide (10 and 14) by 2
5 |  5,  7, 15   ← Divide by 5
      1,  7,  3  ← The only common
                   factor is 1
```

Multiply all the common factors and the residual factors to find the LCM.

LCM of 10, 14, and 15 = 2 × 5 × 1 × 7 × 3
                      = 210

Step 2: Find the number that must be added to 200 so the sum is 210.

This number is the difference between 210 and 200.

    210 − 200 = 10

So 10 must be added to 200 so the sum is divisible by 10, 14, and 15.

Write or choose the letter of the answer.

1. Which of the following numbers is divisible by 3?

    (a) 555
    (b) 480
    (c) 110
    (d) None of the above

    Answer: _____

2. Find the GCF of the following numbers.

    $\lfloor$ 40, 75, 80 _____

    Answer: _____

3. What is the smallest number that must be added to 160 so the sum is divisible by 6, 18, and 20?

    Answer: _____

4. Which of the following numbers is divisible by 8?

    (a) 320
    (b) 288
    (c) 437
    (d) None of the above

    Answer: _____

5. What is the smallest number that must be added to 200 so the sum is divisible by 12, 15, and 16?

    Answer: _____

6. Find the LCM of the following numbers.

    $\lfloor$ 35, 60, 75 _____

    Answer: _____

7. Which of the following numbers is divisible by 6?

    (a) 500
    (b) 666
    (c) 121
    (d) None of the above

    Answer: _____

8. Find the LCM of the following numbers.

    $\lfloor$ 60, 45, 70 _____

    Answer: _____

9. Find the GCF of the following numbers.

    $\lfloor$ 135, 360, 225 _____

    Answer: _____

10. What is the smallest number that must be added to 350 so the sum is divisible by 15, 80, and 90?

    Answer: _____

4.4 GCF and LCM Basics - 2 (**)

Example 1:

Find the GCF of the following numbers.

$$30, 6, 42$$

Solution:

The greatest common factor (GCF) can be found using the following steps.

- Divide all the numbers by a common factor.

- Continue dividing by common factors until there is no factor of all the numbers left.

- Multiply only the common factors to find the GCF.

$$
\begin{array}{r|l}
2 & 30, 6, 42 \quad \leftarrow \text{ Divide by 2} \\
\hline
3 & 15, 3, 21 \quad \leftarrow \text{ Divide by 3} \\
\hline
& 5, 1, 7 \quad \leftarrow \text{ No more common} \\
& \qquad\qquad\quad \text{ factors}
\end{array}
$$

Multiply the common factors 2 and 3 to find the GCF.

GCF of 30, 6, and 42 = $2 \times 3 = 6$

Note that the GCF of 30, 6, and 42 is the largest number that divides these numbers.

Example 2:

Sara is packing fruit baskets. She has 21 kilograms of apples, 35 kilograms of mangoes, and 56 kilograms of grapes. All baskets must have the same number of each item with no items left over. What is the maximum number of baskets she can pack?

Solution:

Since each item is equally divided and no item is left over, the number of fruit baskets will be the GCF of 21, 35, and 56.

$$
\begin{array}{r|l}
7 & 21, 35, 56 \quad \leftarrow \text{ Divide by 7} \\
\hline
& 3, \ 5, \ 8 \quad\;\; \leftarrow \text{ No more common} \\
& \qquad\qquad\quad\;\; \text{ factors}
\end{array}
$$

Multiply the common factors to find the GCF.

GCF of 21, 35, and 56 = 7

So Sara can make a maximum of 7 fruit baskets.

Write the answer.

1. Manoj is filling some boxes. He has 30 ribbons, 12 headbands, and 42 beads. All boxes must have the same number of each item, with no items left over. What is the maximum number of boxes he can fill?

    Answer: _____ _____
    unit

2. Find the LCM of the following numbers.

    | 56, 72, 100 |

    Answer: _____

3. Lisa is filling some jars. She has 60 red marbles, 72 white marbles, and 93 blue marbles. All jars must have the same number of each color marble, with no marbles left over. What is the maximum number of jars she can fill?

    Answer: _____ _____
    unit

4. Find the GCF of the following numbers.

    | 34, 119, 136 |

    Answer: _____

5. Find the LCM of the following numbers.

    | 5, 2, 12 |

    Answer: _____

6. Lora is making bouquets. She has 55 orchids, 70 roses, and 40 lilies. All bouquets must have the same number of each flower, with no flowers left over. What is the maximum number of bouquets she can make?

    Answer: _____ _____
    unit

7. Find the GCF of the following numbers.

    | 15, 22, 31 |

    Answer: _____

8. Charlie is preparing snack packets for a picnic. He has 32 cakes, 10 cookies, and 14 pieces of candy. All packets must have the same number of each item with no items left over. What is the maximum number of packets?

    Answer: _____ _____
    unit

9. Find the LCM of the following numbers

    | 28, 35, 42 |

    Answer: _____

10. Find the GCF of the following numbers.

    | 40, 5, 82 |

    Answer: _____

4.5 Review of Chapter 4 (**)

Write or choose the letter of the answer.

1. What is the sum of the place values of 3, 4, and 9 in 23,749?

   Answer: _____

2. What is the smallest number with one decimal place that is greater than 10?

   Answer: _____

3. Find the LCM of the following numbers.

   | 56, 21, 88 |

   Answer: _____

4. A number has 2 decimal places. It is greater than 25 and less than 26. The tenths-place digit is 4 less than the hundredths-place digit. If the hundredths-place digit is 9, what is the decimal number?

   Answer: _____

5. Sonal is packing tiffin boxes. She has 18 pieces of sweet, 36 bananas, and 27 eggs. All tiffin boxes must have the same number of each item, with no items left over. What is the maximum number of tiffin boxes she can pack?

   Answer: ____ _____
                           unit

6. What is the sum of the values of 5, 2, and 8 in 57,238?

   Answer: _____

7. What is the largest 2-digit number that is divisible by 6?

   Answer: _____

8. Find the LCM of the following numbers.

   | 14, 7, 35 |

   Answer: _____

9. A number has 2 decimal places. It is greater than 30 and less than 31. The tenths-place digit is 2 less than the hundredths-place digit. If the hundredths-place digit is 6, what is the decimal number?

   Answer: _____

10. What is the largest 4-digit number that is less than 10,000?

   Answer: _____

Write or choose the letter of the answer.

11. Which of the following numbers is divisible by 7?

    (a) 840

    (b) 200

    (c) 490

    (d) 691

               Answer: _____

12. Rohan is packing vegetable baskets. He has 42 potatoes, 68 carrots, and 56 peppers. All baskets must have the same number of each item, with no items left over. What is the maximum number of baskets he can pack?

               Answer: ____ _____
                                      unit

13. What is the difference between the place values of 9 and 5 in 396.58?

               Answer: _____

14. Find the LCM of the following numbers.

    84, 96, 72

               Answer: _____

15. What is the sum of the values of 9, 3, and 6 in 93.86?

               Answer: _____

16. Neha is filling some fish jars. She has 6 green fish, 18 red fish, and 9 blue fish. All jars must have the same number of each color fish, with no fish left over. What is the maximum number of jars she can fill?

               Answer: ____ _____
                                      unit

17. A number has 3 decimal places. It is greater than 17 and less than 18. All digits after the decimal point are 4 more than the thousandths-place digit. If the thousandths-place digit is 5, what is the decimal number?

               Answer: _____

18. What is the sum of the place values of 8 and 4 in 861.74?

               Answer: _____

19. What is the smallest number with two decimal places that is greater than 958?

               Answer: _____

20. What is the largest 4-digit number that is less than 7,517?

               Answer: _____

5. Age Problems

5.1 Basic Word Problems on Age (*)

Example 1:

Jacob is currently 12 years old. How old was he 4 years ago, and how old will he be 7 years from now?

Solution:

As given in the question:

Jacob's current age = 12 years

You can use the following steps to answer the questions:

- To find Jacob's age 4 years ago, subtract 4 years from his current age.

  Jacob's age 4 years ago

  = (Jacob's current age) − 4

  = 12 − 4 = 8 years

- To find his age 7 years from now, add 7 years to his current age.

  Jacob's age 7 years from now

  = (Jacob's current age) + 7

  = 12 + 7 = 19 years

So 4 years ago Jacob was 8 years old, and 7 years from now he will be 19 years old.

Example 2:

Disha was 13 years old in June of 2014. How old will she be in June of 2023, and how old was she in June of 2006?

Solution:

As given in the question:

Disha's age in June of 2014 = 13 years

You can use the following steps to answer the questions:

- Difference between June 2023 and June 2014 = 2023 − 2014

  = 9 years

  To find Disha's age in June of 2023, add 9 years to her age in June of 2014.

  Disha's age in June of 2023

  = (Disha's age in June of 2014) + 9

  = 13 + 9 = 22 years

- Difference between June 2014 and June 2006 = 2014 − 2006

  = 8 years

  To find Disha's age in June of 2006, subtract 8 years from her age in June of 2014.

  Disha's age in June of 2006

  = (Disha's age in June of 2014) − 8

  = 13 − 8 = 5 years

Write the answer.

1. Kevin will be 11 years old in 3 years. How old was he 6 years ago?

    Answer: ____ _____
    unit

6. Mark is currently 14 years old. How old was he 9 years ago?

    Answer: ____ _____
    unit

2. Grace is currently 17 years old. How old will she be 14 years from now?

    Answer: ____ _____
    unit

7. Sofia was 11 years old 4 years ago. How old is she now?

    Answer: ____ _____
    unit

3. Anisha is currently 20 years old. How old was she 11 years ago?

    Answer: ____ _____
    unit

8. Juhi was 16 years old in July of 2009. How old will she be in July of 2015?

    Answer: ____ _____
    unit

4. Karan was 8 years old in April of 2007. How old was he in April of 2001?

    Answer: ____ _____
    unit

9. Niraj is currently 37 years old. How old will he be 14 years from now?

    Answer: ____ _____
    unit

5. Olivia will be 22 years old in May of 2019. How old will she be in May of 2026?

    Answer: ____ _____
    unit

10. Franc was 23 years old in August of 1992. How old will he be in August of 1998?

    Answer: ____ _____
    unit

5.2 Sum of and Difference between Ages (**)

Example 1:

The sum of Rohit's and John's ages is 32. What will the sum of their ages be 7 years from now?

Solution:

As given in the question:

Sum of Rohit's and John's ages = 32 years

You can use the following steps to answer the questions:

- Find the ages 7 years from now

  Rohit's age 7 years from now
  = (Rohit's current age) + 7
  John's age 7 years from now
  = (John's current age) + 7

- Find sum of the ages 7 years from now

  Sum of their ages 7 years from now
  = (Rohit's age 7 years from now)
  + (John's age 7 years from now)
  = (Rohit's current age) + 7
      + (John's current age) + 7
  = (Rohit's current age)
      + (John's current age) + 7 + 7
  = (Sum of Rohit's and John's ages)
                    + 14
  = 32 + 14 = 46 years

So the sum of their ages 7 years from now will be 46 years.

Example 2:

The difference between Saan's and Mia's ages is 4. What will the difference in their ages be 10 years from now?

Solution:

The following information is given:

Difference between Saan's and Mia's current ages = 4 years

You can find the difference in their ages at a different time using the following steps:

Find the difference between their ages 10 years from now.

Saan's age 10 years from now
= (Saan's current age) + 10

Mia's age 10 years from now
= (Mia's current age) + 10

Difference between their ages in 10 years
= (Saan's age 10 years from now)
  – (Mia's age 10 years from now)
= (Saan's current age) + 10
          – (Mia's current age) – 10
= (Saan's current age)
          – (Mia's current age)
= (difference between their current
                              ages)
= 4 years

So the difference in their ages 10 years from now will be 4 years.

Note: The difference between two people's ages always remains the same. This difference is the age the older person was when the younger person was born.

Write the answer.

1. The sum of Eric's and Bob's ages is 45. What will the sum of their ages be 11 years from now?

Answer: ____ _____
unit

2. Roshni's current age is two-sixths of her father's age. How much will Roshni's age increase in 8 years?

Answer: ____ _____
unit

3. How much will the sum of Roshni's and her father's ages in question 2 change in 8 years?

Answer: ____ _____
unit

4. The sum of Max's and Alex's ages is 27. What was the sum of their ages 9 years ago?

Answer: ____ _____
unit

5. The difference between Kavya's and Neha's ages is 10. What will the difference between their ages be 12 years from now?

Answer: ____ _____
unit

6. The sum of Naksh's and Joy's ages is 42. What was the sum of their ages 6 years ago?

Answer: ____ _____
unit

7. The difference between Nikhil's and Brian's ages is 15. What was the difference between their ages 5 years ago?

Answer: ____ _____
unit

8. The sum of Julie's and Alice's ages is 25. What will the sum of their ages be in 7 years?

Answer: ____ _____
unit

9. The difference between Stephen's and Philip's ages is 6. What will the difference between their ages be 20 years from now?

Answer: ____ _____
unit

10. The sum of Pamela's and Angela's ages is 30. What will the sum of their ages be in 10 years?

Answer: ____ _____
unit

5.3 Solving Age Problems at One Time (**)

Example 1:

Paul is presently 2 times as old as Masum. The sum of their ages is 39. What are Paul's and Masum's current ages?

Solution:

You can use a variable (P) for Masum's current age and solve the problem as shown below:

- Define the problem mathematically.

  Masum's current age = P
  Paul's current age

  = 2 times Masum's current age

  = $2P$

- Write the equation.
  Sum of their current ages is 39.

  (Masum's current age)
  + (Paul's current age) = 39

  $P + 2P = 39$
  $3P = 39$

- Solve for P.

  $3P = 39$

  $P = \dfrac{39}{3} = 13$

- Find the ages.

  Masum's current age = P = 13 years

  Paul's current age = $2P$
  = 2 × 13
  = 26 years

Example 2:

Helen is presently 2 times as old as Rian. The difference between their ages is 8. How old will Rian be in 7 years, and how old will Helen be in 7 years?

Solution:

You can use a variable (R) for Rian's current age and solve the problem as shown below:

- Define the problem mathematically.

  Rian's current age = R
  Helen's current age

  = 2 times Rian's current age

  = $2R$

- Write the equation.
  Difference between their current ages is 8.

  (Helen's current age)
  − (Rian's current age) = 8

  $2R − R = 8$
  $R = 8$

- Find the ages.

  Rian's current age = R = 8 years

  Helen's current age = $2R$
  = 2 × 8 = 16 years

- Find their ages in 7 years.

  Rian's age in 7years
  = (Rian's current age) + 7
  = 8 + 7 = 15 years

  Helen's age in 7 years
  = (Helen's current age) + 7
  = 16 + 7 = 23 years

Write the answer.

1. Ben is presently 3 times as old as Daniel. The sum of their ages is 48. What is Daniel's current age?

    Answer: _____ _____
    unit

2. Karan is presently 6 times as old as Kavya. The sum of their ages is 28. What is Karan's current age?

    Answer: _____ _____
    unit

3. Steven is currently 4 times as old as Colin. Six years ago, the difference between their ages was 33. What is Colin's current age?

    Answer: _____ _____
    unit

4. Nikhil is presently 3 times as old as Jyoti. The difference between their ages is 10. How old will Nikhil be in 4 years?

    Answer: _____ _____
    unit

5. Sarah is presently 4 times as old as Patricia. The sum of their ages is 40. What is Sarah's current age?

    Answer: _____ _____
    unit

6. Carlson was 5 times as old as Stella 5 years ago. The sum of their ages 5 years ago was 72. How old was Carlson?

    Answer: _____ _____
    unit

7. Suman is currently 2 times as old as Ronak. Six years ago, the difference between their ages was 10. What will Ronak's age be 5 years from now?

    Answer: _____ _____
    unit

8. In May of 2006, Maria was 2 times as old as her brother. The product of their ages was 50. What was Maria's brother's age in May 2006?

    Answer: _____ _____
    unit

5.4 Solving Age Problems at Two Different Times (**)

Example 1:

Watson is presently 3 times as old as Nil. The sum of their ages is 40. How old was Watson 8 years ago, and how old was Nil 5 years ago?

Solution:

You can use a variable (X) for Nil's current age and solve the problem as shown below:

- Define the problem mathematically.

  Nil's current age = X
  Watson's current age
       = 3 times Nil's current age
       = $3X$

  Nil's age 5 years ago = $X - 5$

  Watson's age 8 years ago = $3X - 8$

- Write the equation.
  Sum of their current ages is 40.

  (Nil's current age)
       + (Watson's current age) = 40

  $X + 3X = 40$
  $4X = 40$

- Solve for X.

  $4X = 40$

  $X = \dfrac{40}{4} = 10$

- Find the ages

  Watson's age 8 years ago = $3X - 8$
           = $(3 \times 10) - 8$
           = $30 - 8 = 22$ years

  Nil's age 5 years ago = $X - 5$
           = $10 - 5$
           = 5 years

Example 2:

Nikhil is presently 2 times as old as Neeraj. Three years from now, the sum of their ages will be 24. What are Nikhil's and Neeraj's current ages?

Solution:

You can use a variable (X) for Neeraj's current age and solve the problem as shown below:

- Define the problem mathematically.

  Neeraj's current age = X
  Nikhil's current age
       = 2 times Neeraj's current age
       = $2X$

  Neeraj's age in 3 years = $X + 3$

  Nikhil's age in 3 years = $2X + 3$

- Write the equation.

  3 years from now, sum of their ages
           = 24

  (Neeraj's age in 3 years)
       + (Nikhil's age in 3 years) = 24
  $X + 3 + 2X + 3 = 24$
  $3X + 6 = 24$

- Solve for X.

  $3X + 6 - 6 = 24 - 6$
  $3X = 18$

  $X = \dfrac{18}{3} = 6$

- Find the ages

  Neeraj's current age = X = 6 years

  Nikhil's current age = $2X$
           = 2×6
           = 12 years

Write the answer.

1. Ritwik is presently 2 times as old as Subham. Six years from now, the sum of their ages will be 42. What is Subham's current age?

    Answer: ____ _____
    unit

2. Andy is presently 3 times as old as Max. The sum of their ages is 36. How old will Andy be 5 years from now?

    Answer: ____ _____
    unit

3. Sweta is presently 4 years older than Roni. If Roni is 22 years old now, how old will Sweta be in 9 years?

    Answer: ____ _____
    unit

4. Lora is presently 8 years younger than Sarika. The sum of their present ages is 28. How old was Sarika 3 years ago?

    Answer: ____ _____
    unit

5. Aman is presently 4 years older than Alex. The sum of their present ages is 44. How old will Aman be in 7 years?

    Answer: ____ _____
    unit

6. Neha is presently 4 times as old as Mohali. Five years from now, the sum of their ages will be 50. How old is Neha now?

    Answer: ____ _____
    unit

7. Arnav is presently 12 years older than Mark. If Mark is 16 years old now, how old was Arnav 5 years ago?

    Answer: ____ _____
    unit

8. Rupak is presently 3 times as old as Suhani. Two years from now, the sum of their ages will be 52. What is Rupak's current age?

    Answer: ____ _____
    unit

5.5 Review of Chapter 5 (**)

Write the answer.

1. Nancy will be 21 years old in 5 years. How old was she 6 years ago?

   Answer: _____ _____
   unit

2. Ava is presently 7 years younger than Georgia. The sum of their present ages is 23. How old will Ava be in 4 years?

   Answer: _____ _____
   unit

3. Carlos was 5 times as old as his son 8 years ago. The sum of their ages 8 years ago was 42. What is his son's current age?

   Answer: _____ _____
   unit

4. The difference between Emily's and Angela's current ages is 3. The sum of their ages 4 years ago was 19. What was Emily's age 2 years ago?

   Answer: _____ _____
   unit

5. The sum of Craig's and Eric's current ages is 30. What will the sum of their ages be in 7 years?

   Answer: _____ _____
   unit

6. Carter is presently 3 times as old as Evelyn. The sum of their ages is 24. What is Carter's current age?

   Answer: _____ _____
   unit

Write the answer.

7. Olivia is presently 5 years older than Carl. If Carl is 17 years old now, how old will Olivia be in 3 years?

10. Ricky is currently 3 times as old as Nihal. Four years ago, the difference between their ages was 20. What is Nihal's current age?

Answer: _____ _____
unit

Answer: _____ _____
unit

8. The sum of Diego's and Jasmine's current ages is 45. What was the sum of their ages 10 years ago?

11. The difference between Caroline's and Amy's ages is 20. What was the difference between their ages 8 years ago?

Answer: _____ _____
unit

Answer: _____ _____
unit

9. In July of 2010, Franc was 2 times as old as his sister. The product of their ages was 32. What was Franc's age in July 2010?

12. Ben is currently 15 years old. How old will he be 9 years from now?

Answer: _____ _____
unit

Answer: _____ _____
unit

6. Time and Distance Problems

6.1 Find Speed Given Distance and Time (**)

Example 1:

Find the unit rate.

15 pages in 5 minutes

Solution:

Find the unit rate by calculating pages in 1 minute or pages per minute.

$$\text{Unit rate} = \frac{15 \text{ pages}}{5 \text{ minutes}}$$

$$= \frac{15}{5} \frac{\text{pages}}{\text{minute}}$$

$$= \frac{3}{1} \frac{\text{pages}}{\text{minute}}$$

$$= 3 \text{ pages per minute}$$

So the unit rate is 3 pages per minute.

Example 2:

If a car travels 420 kilometers in 4 hours, what is the average speed in kilometers per hour?

Solution:

The following information is given:

Distance = 420 kilometers

Time = 4 hours

Find the average speed in kilometers per hour.

$$\text{Average speed} = \frac{\text{distance}}{\text{time}}$$

$$= \frac{420 \text{ kilometers}}{4 \text{ hours}}$$

$$= \frac{420}{4} \frac{\text{kilometers}}{\text{hour}}$$

$$= \frac{105}{1} \frac{\text{kilometers}}{\text{hour}}$$

$$= 105 \text{ kilometers per hour}$$

So the average speed of the car is 105 kilometers per hour.

Write the answer.

1. A man can run 220 meters in 5 minutes. What is his average speed in meters per minute?

2. Find the unit rate for the following expression:

    21 pastries in 7 boxes

Answer: _____ _____
              unit

Answer: _____ _____
              unit

Write the answer.

3. Find the unit rate for the following expression:

    120 biscuits for 10 students

    Answer: ____ _____
    unit

4. Find the unit rate for the following expression:

    81 people in 9 vans

    Answer: ____ _____
    unit

5. A tiger ran 390 miles in 6 hours. What was its average speed in miles per hour?

    Answer: ____ _____
    unit

6. A sailfish traveled 432 kilometers in 4 hours. What was the average speed in kilometers per hour?

    Answer: ____ _____
    unit

7. If a racing bike can travel 527 kilometers in 4 hours and 15 minutes, what is the average speed in kilometers per hour?

    Answer: ____ _____
    unit

8. Find the average speed in miles per minute:

    760 miles in 6 hours and 20 minutes

    Answer: ____ _____
    unit

9. Find the unit rate for the following expression:

    780 dollars for 52 tickets

    Answer: ____ _____
    unit

10. Find the average speed in kilometers per hour:

    897 kilometers in 13 hours

    Answer: ____ _____
    unit

11. Find the unit rate for the following expression:

    9 tasks in 3 hours

    Answer: ____ _____
    unit

12. A train travels 756 kilometers in 9 hours. What is the average speed in kilometers per hour?

    Answer: ____ _____
    unit

6.2 Find Distance Given Speed and Time (**)

Example 1:

Sam and his friends drove 6 hours during a trip. They drove at a speed of 50 miles per hour for the first half of the trip and at 40 miles per hour for the second half of the trip. How far did they drive during the entire trip?

Solution:

The following information is given:

Total time for the trip = 6 hours
Speed in first half = 50 miles per hour
Speed in second half = 40 miles per hour

You can solve this problem in three steps.

- Find the distance traveled in the first half of the trip.

  Time for first half = $\dfrac{6 \text{ hours}}{2}$ = 3 hours

  Distance traveled in first half
  = (speed in first half)
      × (time for first half)
  = 50 × 3 = 150 miles

- Find the distance traveled in the second half of the trip.

  Time for second half = $\dfrac{6 \text{ hours}}{2}$
  = 3 hours

  Distance traveled in second half
  = (speed in second half)
      × (time for second half)
  = 40 × 3 = 120 miles

- Find the total distance.

  Total distance
  = (distance traveled in first half)
      + (distance traveled in second half)
  = 150 + 120 = 270 miles

So they drove 270 miles during the entire trip.

Example 2:

Find the distance a car traveled in the following situation:
  Average speed of the car
              = 67 kilometers per hour
  Time of travel = 4 hours

Solution:

The following information is given:
  Average speed = 67 kilometers per hour
  Time = 4 hours

You can find the distance using the following formula:

  Distance = (average speed) × (time)
          = 67 × 4
          = 268 kilometers

So the car traveled a distance of 268 kilometers in 4 hours.

Name _____

Write the answer.

1. Mr. Clark and his family traveled for 4 hours. They traveled at a speed of 75 kilometers per hour for the first half of the trip and at 80 kilometers per hour for the second half. How far did they travel?

   Answer: ____ _____
   unit

2. Find the distance a bicycle traveled in the following situation:
   Average speed of the bicycle
   = 12 kilometers per hour
   Time of travel = 3 hours

   Answer: ____ _____
   unit

3. Car A and car B started from the same starting point and traveled at different speeds. Car A traveled at a speed of 44 kilometers per hour, and car B traveled at a speed of 35 kilometers per hour. How far apart will they be after 3 hours?

   Answer: ____ _____
   unit

4. Find the distance a dog ran in the following situation:
   Average speed of the dog
   = 15 miles per hour
   Time of travel = 5 hours

   Answer: ____ _____
   unit

5. Andy and David started from the same starting line and traveled at different speeds. Andy traveled at a speed of 15 miles per hour, and David traveled at a speed of 12 miles per hour. How far apart will they be after 5 hours?

   Answer: ____ _____
   unit

6. A group of girls traveled 8 hours during an educational tour. They traveled at a speed of 46 miles per hour for the first half of the trip and at 59 miles per hour for the second half of the trip. How far did they travel during the entire tour?

   Answer: ____ _____
   unit

7. Find the distance a man climbed in the following situation:
   Average speed of the man
   = 2 kilometers per hour
   Time of travel = 4 hours

   Answer: ____ _____
   unit

8. Three women took a vacation together. They traveled 12 hours during the trip. They traveled at a speed of 86 kilometers per hour for the first half of the trip and at 43 kilometers per hour for the second half of the trip. How far did they travel during the entire trip?

   Answer: ____ _____
   unit

6.3 Find Time Given Speed and Distance (**)

Example 1:

Lisa and Lora started a race from two opposite sides, side A and side B. Lisa started from side A and ran toward side B. Lora started from side B at the same time and ran toward side A. Lisa ran 1,150 meters at a speed of 50 meters per minute before she saw Lora. How many minutes had Lora run before she saw Lisa?

Solution:

The following information is given:

Distance Lisa ran before she saw Lora = 1,150 meters

Lisa's speed = 50 meters per minute

You can find the time using the following formula.

$$\text{Time} = \frac{\text{distance}}{\text{speed}}$$

$$= \frac{1150 \text{ meters}}{50 \text{ meters per minute}}$$

$$= \frac{1150}{50} \frac{\text{meters}}{\text{meters per minute}}$$

$$= 23 \text{ minutes}$$

Both of them started at the same time. That means they ran the same amount of time before they saw each other.

So Lora also ran for 23 minutes before she saw Lisa.

Example 2:

Mr. White traveled 280 miles to reach his friend's house. He traveled at a speed of 70 miles per hour for half of the distance and at 35 miles per hour for the remaining distance. How long did he take to travel 280 miles?

Solution:

The following information is given:

Total distance = 280 miles

Speed for first half of the distance = 70 miles per hour

Speed for the remaining distance = 35 miles per hour

Using the above information, you can find the answer:

Distance for first half = 280 ÷ 2 = 140 miles

$$\text{Time} = \frac{\text{distance}}{\text{speed}} = \frac{140 \text{ miles}}{70 \text{ miles per hour}}$$

$$= \frac{140}{70} \frac{\text{miles}}{\text{miles per hour}} = 2 \text{ hours}$$

Distance for second half = 280 − 140 = 140 miles

$$\text{Time} = \frac{\text{distance}}{\text{speed}} = \frac{140 \text{ miles}}{35 \text{ miles per hour}}$$

$$= \frac{140}{35} \frac{\text{miles}}{\text{miles per hour}} = 4 \text{ hours}$$

Total time = time for first half of distance + time for second half of distance = 2 hours + 4 hours = 6 hours

So Mr. White took 6 hours to travel 280 miles.

Write the answer.

1. Jack participated in a 200-meter walking race. His average speed was 20 meters per minute. How long did he take to complete the race?

Answer: ____ _____
unit

2. Mr. Johnson traveled 560 kilometers to reach a tournament. He traveled at a speed of 140 kilometers per hour for half of the distance, took a break of 1 hour, and traveled at a speed of 70 kilometers per hour for the remaining distance. How long did it take to complete the 560-kilometer trip, including break time?

Answer: ____ _____
unit

3. Train A started from Lancaster and traveled toward Visalia. Train B started from Visalia at the same time and traveled toward Lancaster. Train A traveled for 276 miles at a speed of 46 miles per hour before it saw train B. How many hours had train B traveled before it saw train A?

Answer: ____ _____
unit

4. An archery arrow traveled 450 meters to hit a target. Its average speed was 90 meters per second. How many seconds did the arrow travel before hitting the target?

Answer: ___ _____
unit

5. A loaded truck had to travel 140 miles. It completed 120 miles at a speed of 40 miles per hour. Due to a traffic jam, it completed the remaining distance at a speed of 20 miles per hour. What was the truck's average speed during the entire trip?

Answer: ____ _____
unit

6. Andy had to travel 180 miles by bus to visit a tourist spot. He completed the first 120 miles at a speed of 40 miles per hour. Due to some engine problems, he completed the remaining distance at a speed of 30 miles per hour. What was the average speed during the entire trip?

Answer: ____ _____
unit

7. Mrs. Lee walks for 480 meters every morning. Her average speed is 15 meters per minute. How long does she walk every morning?

Answer: ____ _____
unit

8. Alka walks 1,600 meters every day for school. Her average speed is 20 meters per minute. How long does Alka walk every day?

Answer: ____ _____
unit

6.4 Rate and Speed Problems (**)

Example 1:

Mr. Martin paid $276.00 to buy four cell phones. He told his wife that the unit rate for his purchase was $69.00 phone/dollar. Identify the error and choose the correct unit rate.

(a) $69.00 phone per dollar
(b) $69.00 per phone
(c) $69.00 per 4 phones
(d) $69.00 phone/dollar

Solution:

The unit rate is the price for 1 phone. This is also called the price per phone.

The following information is given:

Amount paid = $276.00
Number of phones purchased = 4

$$\text{Unit rate} = \frac{\$276.00}{4 \text{ phones}}$$

$$= \frac{\$69.00}{1 \text{ phone}}$$

$$= \$69.00 \text{ per mobile}$$

So Mr. Martin paid $69.00 per phone, and the correct answer is (b).

Example 2:

One cookie packet contains 18 cookies and serves 6 children. Write the number of cookies per child as a unit rate, and find how many packets are needed to serve 12 children.

Solution:

The following information is given.

Total number of children = 6
Number of cookies in 1 packet =18

$$\text{Unit rate} = \frac{18 \text{ cookies}}{6 \text{ children}} = \frac{18}{6} \frac{\text{cookies}}{\text{child}}$$

$$= 3 \text{ cookies per child}$$

Number of cookies served to 12 children
= (total number of children)
    × (number of cookies per child)
= 12 × 3 = 36 cookies

Number of packets needed
= (number of cookies served)
    ÷ (number of cookies in 1 packet)
= 36 ÷ 18 = 2 packets
So you need 2 packets to serve 12 children.

Write or choose the letter of the answer.

1. Jyoti has the following two options to buy burgers. Which is the better buy?

   (a) 18 burgers for $168.33
   (b) 13 burgers for $116.48

   Answer: _____ _____
                         unit

2. A fruit seller sells bananas in two ways: 1 dozen bananas for $4.80 or 15 bananas for $5.25. Which is the better buy?

   (a) 1 dozen bananas
   (b) 15 bananas

   Answer: _____ _____
                         unit

Write or choose the letter of the answer.

3. There are 64 players going to a hockey tournament. The organizing committee wants leaders for the event and needs to have a ratio less than or equal to 16 players to 1 leader. What is the minimum number of leaders needed?

Answer: _____ _____
unit

4. One chocolate bar contains 12 pieces of chocolates and can serve 4 kids. Find how many bars are needed to serve 48 kids.

Answer: _____ _____
unit

5. Lucy has the following two options to buy ice-cream cones. Which is the better buy?

    (a) 8 ice-cream cones for $16.80
    (b) 5 ice-cream cones for $11.50

Answer: _____ _____
unit

6. David paid $52.80 to buy 8 books. He told his friend that the unit rate for his purchase was $6.60 books/dollar. Identify the error and choose the correct unit rate.

    (a) $6.60 per 8 books
    (b) $6.60 book/dollar
    (c) $6.60 book per dollar
    (d) $6.60 per book

Answer: _____ _____
unit

7. Kiran spent $48.00 to buy 4 sets of headphones. She told her uncle that she paid $12.00 for each set. Identify the error, and choose the correct unit rate.

    (a) $12.00 per set
    (b) $38.00 per set
    (c) $14.00 per set
    (d) $28.00 per set

Answer: _____ _____
unit

8. One bag contains 9 pastries and can serve 3 children. Find how many bags are needed to serve 18 children.

Answer: _____ _____
unit

9. Which option is the better buy: 7 pounds of nuts for $14.07 or 8 pounds of nuts for $16.24?

    (a) 7 pounds of nuts
    (b) 8 pounds of nuts

Answer: _____ _____
unit

10. A camp donates clothes to less fortunate people. One bag contains 25 pieces of clothing for 5 people. Write the number of clothes per person as a unit rate. Find how many bags are needed for 125 people.

Answer: _____ _____
unit

6.5 Advanced Speed and Distance Problems 1 (***)

Example 1:

Jack and Ron are traveling in different cars from the same place. Jack is 120 kilometers ahead of Ron. If Jack is traveling at 64 kilometers per hour and Ron is traveling at 70 kilometers per hour, how long will it take for Ron to overtake Jack?

Solution:

The following information is given:

Jack's speed = 64 kilometers per hour

Ron's speed = 70 kilometers per hour

Distance between Jack and Ron = 120 kilometers

You can use the following steps to answer the questions.

- Find the relative speed between Jack and Ron.

Relative speed between Ron and Jack
= (Ron's speed) – (Jack's speed)
= 70 – 64
= 6 kilometers per hour

- Find the time it takes Ron to overtake Jack.

Time Ron takes to overtake Jack

$$= \frac{\text{Distance between Jack and Ron}}{\text{Relative speed between Jack and Ron}}$$

$$= \frac{120}{6}$$

= 20 hours

So Ron will take 20 hours to overtake Jack.

Example 2:

The speed of the water current in a river is 6 kilometers per hour. A ship can travel at 9 kilometers per hour in still water. If the ship is traveling upstream, how much time will it take to travel a distance of 15 kilometers against the riverbank?

Solution:

The following information is given:
Distance = 15 kilometers
Speed of water current
= 6 kilometers per hour
Speed of the ship in still water
= 9 kilometers per hour

Find the time taken to travel 15 kilometers upstream along the riverbank.

When the ship is traveling upstream, the speed of the ship along the riverbank is the difference between the speed of the ship in still water and the speed of the water current.

Boat speed against riverbank
= (speed of the ship in still water)
– (speed of water current)
= 9 – 6 = 3 kilometers per hour

Time = (distance) ÷ (boat speed against riverbank)

$$= \frac{15}{3} = 5 \text{ hours}$$

So it will take the ship 5 hours to travel a distance of 15 kilometers against the riverbank.

Write the answer.

1. Two cars are traveling from one location to another. Car A is 240 miles ahead of car B. If car A is traveling at 52 miles per hour and car B is traveling at 60 miles per hour, how long will it take for car B to overtake car A?

   Answer: _____ _____
   unit

2. Allen ran the first 4 kilometers of a 7-kilometer race at a speed of 200 meters per minute. Then he ran the last 3 kilometers at a speed of 100 meters per minute. What was his average speed for the entire race?

   Answer: _____ _____
   unit

3. The speed of the water current in a river is 5 kilometers per hour. A shark can swim at 45 kilometers per hour in still water. If the shark is swimming upstream, how much time will it take to swim a distance of 360 kilometers against the river?

   Answer: _____ _____
   unit

4. The speed of the water current in a river is 5 kilometers per hour. A boat can travel at 7 kilometers per hour in still water. If the boat is traveling downstream, how much time will it take to travel a distance of 84 kilometers along the riverbank?

   Answer: _____ _____
   unit

5. A truck traveled the first 5,400 meters of an 8,400-meter-long bridge at a speed of 600 meters per minute. Then the truck traveled the last 3,000 meters at a speed of 250 meters per minute. What was its average speed for the entire bridge?

   Answer: _____ _____
   unit

6. The speed of the water current in a river is 3.5 miles per hour. A ship can travel at 8.5 miles per hour in still water. If the ship is traveling upstream, how much time will it take to travel a distance of 50 miles against the riverbank?

   Answer: _____ _____
   unit

7. Two bikers, John and Bill, are riding bikes along the same road. John is 12 miles ahead of Bill. If John is riding at 34 miles per hour and Bill is riding at 32 miles per hour, how long will it take for Bill to overtake John?

   Answer: _____ _____
   unit

8. The speed of the water current in a river is 3 miles per hour. A dolphin can swim at 33 miles per hour in still water. If the dolphin is swimming downstream, how many hours will it take to swim a distance of 144 miles along the coast?

   Answer: _____ _____
   unit

6.6 Review of Chapter 6 (**)

Write or choose the letter of the answer.

1. Find the average speed in miles per minute.

    870 miles in 7 hours and 15 minutes

    Answer: _____ _____
                     unit

2. A horse ran 109.6 kilometers in 2 hours. What was the average speed in kilometers per hour?

    Answer: _____ _____
                     unit

3. Find the unit rate for the following expression:

    42 chairs in 7 days

    Answer: _____ _____
                     unit

4. Eight people traveled 6 hours during a tour. They traveled at a speed of 41 miles per hour for the first half of the trip and at 53 miles per hour for the second half of the trip. How far did they travel during the entire tour?

    Answer: _____ _____
                     unit

5. Find the distance a car traveled in the following situation:
    Average speed of the car
            = 56 miles per hour
    Time of travel = 4 hours

    Answer: _____ _____
                     unit

6. Find the distance a tiger ran in the following situation:
    Average speed of the tiger
            = 85 kilometers per hour
    Time of travel = 3 hours

    Answer: _____ _____
                     unit

7. One box contains 12 pencils and can be given to 6 children. Determine how many pencil boxes are needed for 24 children.

    Answer: _____ _____
                     unit

8. Kavya has the following two options to buy pens. Which is the better buy?

    (a) 9 pens for $9.18
    (b) 7 pens for $7.21

    Answer: _____ _____
                     unit

9. Manoj has to travel 80 miles by car to visit a construction site. He completed the first 50 miles at a speed of 25 miles per hour. Due to engine problem, he completed the remaining distance at a speed of 15 miles per hour. What was the average speed during the entire trip?

    Answer: _____ _____
                     unit

Write or choose the letter of the answer.

10. Jyoti has to travel 300 kilometers. She completed the first 120 kilometers at a speed of 60 kilometers per hour. Due to heavy rain, she completed the remaining distance at a speed of 45 kilometers per hour. What was her average speed for the entire trip?

Answer: ____ _____
   unit

11. A black marlin and a sailfish moved from the shore and swam at different speeds. The black marlin swam at a speed of 56 miles per hour, and the sailfish swam at a speed of 53 miles per hour. How far apart will they be after 4 hours?

Answer: ____ _____
   unit

12. Rob spent $60.00 to buy 2 books. He told his friend that he paid $30.00 for each book. Identify the error, and choose the correct unit rate.

    (a) $45.00 per book
    (b) $120.00 per book
    (c) $30.00 per book
    (d) $17.00 per book

Answer: ____ _____
   unit

13. If a racing car can travel 420 kilometers in 2 hours and 20 minutes, what is the average speed in kilometers per minute?

Answer: ____ _____
   unit

14. Mr. Peterson traveled with his friends for 2 hours. They traveled at a speed of 65 kilometers per hour for the first half of the trip and at 53 kilometers per hour for the second half. How far did they travel?

Answer: ____ _____
   unit

15. Two buses are traveling from one location to another. Bus A is 120 miles ahead of bus B. If bus A is traveling at 26 miles per hour and bus B is traveling at 30 miles per hour, how long will it take for bus B to overtake bus A?

Answer: ____ _____
   unit

16. Lisa has two options for buying sugar: 5 kilograms of sugar for $15.05 or 6 kilograms of sugar for $18.00. Which is the better buy?

    (a) 5 kilograms of sugar
    (b) 6 kilograms of sugar

Answer: ____ _____
   unit

7. Money Problems

7.1 Concept of Simple Interest (*)

Example 1:

Nancy deposited $500.00 in a bank account that paid a simple interest rate of 5 percent per year for a period of 24 months.

   (a) How much is the principal?

   (b) What is the annual interest rate as a percentage?

   (c) What is the time in years?

Solution:

You can find the answers using the following definitions.

   (a) The principal is the amount of money deposited, borrowed, or invested.

Amount Nancy deposited
$$= \$500.00$$

So the principal is $500.00.

   (b) The annual interest rate is the principle paid or earned per year. Simple interest given by the bank per year = 5 percent

So the annual interest rate is 5 percent.

   (c) Time period = 24 months

$$= \frac{24}{12} \text{ years}$$

$$= 2 \text{ years}$$

So the time period is 2 years.

Example 2:

Jenny deposited some money in her savings account and knew her balance at the end of the year. She can find the interest by _____ the principal from the balance.

   (a) adding

   (b) multiplying

   (c) dividing

   (d) subtracting

Solution:

The interest is the difference between the balance and the principal. So Jenny needs to subtract the principal from the balance to find the interest.

Balance = principal + interest
Interest = balance − principal

So the answer is (d).

Name _____

Write or choose the letter of the answer.

1. Rishi borrowed some money from his uncle and knew his interest at the end of the year. He can find the balance by _____ the principal with the interest.

   (a) adding
   (b) dividing
   (c) multiplying
   (d) subtracting

   Answer: _____

2. What is the formula for calculating simple interest?

   (a) $I = P + r + t$
   (b) $I = \dfrac{PRt}{100}$
   (c) $I = P - r - t$
   (d) $I = Prt + Prt$

   Here,

   I = the interest

   P = the principal

   r = the annual interest rate written as a decimal

   t = the time in years

   Answer: _____

3. When you borrow money from a bank, would you want a low or a high interest rate?

   (a) Low interest rate
   (b) High interest rate
   (c) Does not matter if the interest rate is low or high

   Answer: _____

4. When you invest money in a stock, would you want a low or a high interest rate?

   (a) Low interest rate
   (b) High interest rate
   (c) Does not matter if the interest rate is low or high

   Answer: _____

5. Lisa deposited $1,200.00 in a bank account that will earn a simple annual interest rate of 4.5 percent for a period of 36 months. What is the time in years?

   Answer: ____ _____
   unit

6. What is the formula for calculating the principal?

   (a) $P = B \times I$
   (b) $P = B - I$
   (c) $P = B \div I$
   (d) $P = B + I$

   Here,

   B = the balance

   I = the interest

   P = the principal

   Answer: _____

7. Emi invested $2,500.00 in a mutual fund that will earn a simple annual interest rate of 8 percent for a period of 20 months. How much is the principal?

   Answer: _____

78

7.2 Problems in Simple interest (**)

Example 1:

Simon deposited an amount of $6,000.00 in his bank account. The bank will give him 4 percent annual interest. If he has to withdraw the balance after 3 years, how much will he have to withdraw?

Solution:

The following information is given:

Principal (P) = $6,000.00
Interest rate (R) = 4 percent
Time (t) = 3 years

Calculate the interest (I) by using the following formula:

$$I = \frac{PRt}{100}$$

$$= \frac{6,000 \times 4 \times 3}{100}$$

$$= 60 \times 4 \times 3 = \$720.00$$

Find the balance (B) by adding principal and interest.

$$B = P + I$$
$$= \$6,000.00 + \$720.00$$
$$= \$6,720.00$$

So Simon will withdraw $6,720.00 after 3 years.

Example 2:

Lucy deposited $1,000.00 in a savings account that earns 3.5 percent simple annual interest. Her friend deposited $1,500.00 in a mutual fund that earns 6.5 percent simple annual interest. How much more interest will her friend have earned at the end of 4 years?

Solution:

Find the interest Lucy earned.

Principal (P) = $1,000.00
Interest rate (R) = 3.5 percent
Time (t) = 4 years

$$I = \frac{PRt}{100} = \frac{1,000 \times 3.5 \times 4}{100}$$

$$= 10 \times 3.5 \times 4$$

$$= \$140.00$$

Find the interest her friend earned.

Principal (P) = $1,500.00
Interest rate (R) = 6.5 percent
Time (t) = 4 years

$$I = \frac{PRt}{100} = \frac{1,500 \times 6.5 \times 4}{100}$$

$$= 15 \times 6.5 \times 4$$

$$= \$390.00$$

Difference between their interest amounts
$$= \$390.00 - \$140.00 = \$250.00$$

So Lucy's friend will earn $250.00 more interest than Lucy will.

Write the answer.

1. What is the total interest received on an investment of $8,900.00 at an interest rate of 5.5 percent per year in 4 years?

   Answer: _____

2. Kapil deposited $2,200.00 in a savings account that earns 5 percent simple annual interest. His sister deposited $3,000.00 in a mutual fund that earns 7.2 percent simple annual interest. How much more interest will his sister have earned at the end of 6 years?

   Answer: _____

3. Julie deposited $9,000.00 in her bank account. The bank will give her 3 percent annual interest. If she has to withdraw the balance after 7 years, how much will she withdraw?

   Answer: _____

4. What is the total interest received on an investment of $8,200.00 at an interest rate of 5 percent per year in 4 years?

   Answer: _____

5. Jenny took a loan of $6,900.00 from the bank to buy a motorcycle. The bank charged her 4 percent simple annual interest. If she has to pay back the loan after 2 years, how much will she have to pay back?

   Answer: _____

6. What is the total interest earned on a $9,300.00 deposit at an interest rate of 6.4 percent per year in 5 years?

   Answer: _____

7. Jack deposited $700.00 in an investment that earns 2 percent simple annual interest. Bob deposited $950.00 in an investment that earns 3 percent simple annual interest. How much more interest will Bob earn at the end of 5 years?

   Answer: _____

8. What is the total interest earned on a $4,600.00 deposit at an interest rate of 3.3 percent per year in 5 years?

   Answer: _____

7.3 Find the Annual Interest Rate (**)

Example 1:

Nikhil deposited $6,000.00 in his bank account. At the end of 3 years, his bank balance had gone up by $1,755.00. Find the simple annual interest rate.

Example 2:

An investment of $20,000.00 earns simple interest and becomes $22,800.00 after 4 years. What is the simple annual interest rate?

Solution:

The following information is given:

Principal (P) = $6,000.00
Interest (I) = $1,755.00
Time (t) = 3 years

Use these values in the simple interest formula to find R.

$$I = \frac{PRt}{100} \quad \Rightarrow \quad R = \frac{I \times 100}{Pt}$$

$$R = \frac{I \times 100}{Pt}$$

$$= \frac{1{,}755 \times 100}{6{,}000 \times 3}$$

$$= \frac{1{,}755}{60 \times 3} = 9.75\%$$

So the simple annual interest rate is 9.75 percent.

Solution:

The following information is given:

Principal (P) = $20,000.00
Interest (I) = balance − principal
 = $22,800.00 − $20,000.00
 = $2,800.00
Time (t) = 4 years

Use these values in the simple interest formula to find R.

$$I = \frac{PRt}{100} \quad \Rightarrow \quad R = \frac{I \times 100}{Pt}$$

$$R = \frac{I \times 100}{Pt}$$

$$= \frac{2{,}800 \times 100}{20{,}000 \times 4}$$

$$= \frac{28}{8}$$

$$= 3.5\%$$

So the simple annual interest rate is 3.5 percent.

Write the answer.

1. A principal amount of $3,000.00 earns simple interest and becomes $3,450.00 in 5 years. What is the simple annual interest rate?

2. Arushi deposited $4,000.00 in her bank account. At the end of 4 years, her bank balance had gone up by $640.00. Find the simple annual interest rate.

Answer: _____

Answer: _____

Write the answer.

3. A deposit of $7,000.00 earns simple interest and becomes $7,420.00 after 2 years. What is the simple annual interest rate?

Answer: _____

4. Use the simple interest formula $\left(I = \dfrac{PRt}{100}\right)$ to find the unknown quantity.

$I = \$840.00$
$P = \$4,200.00$
$R = ?$ percent
$t = 4$ years

Answer: _____

5. Disha deposited $1,000.00 in her bank account. At the end of 8 years, her bank balance had gone up by $200.00. Find the simple annual interest rate.

Answer: _____

6. Nikhil invested $1,500.00 in a mutual fund. At the end of 3 years, his balance in the mutual fund had gone up by $180.00. Find the simple annual interest rate.

Answer: _____

7. Julie deposited $7,000.00 in her bank account. At the end of 5 years, her bank balance had gone up by $700.00. Find the simple annual interest rate.

Answer: _____

8. A principal amount of $1,200.00 earns simple interest and becomes $1,800.00 in 5 years. What is the simple annual interest rate?

Answer: _____

9. Jay invested $2,200.00 in a mutual fund. At the end of 6 years, his balance had gone up by $528.00. Find the simple annual interest rate.

Answer: _____

10. Use the simple interest formula $\left(I = \dfrac{PRt}{100}\right)$ to find the unknown quantity.

$I = \$45.00$
$P = \$500.00$
$R = ?$ percent
$t = 4$ years

Answer: _____

Name _____

7.4 Find the Time Period in Problems with Simple Interest (**)

Example 1:

Kavya invested $2,500.00 in a business plan. She received a total of $2,700.00 after several years. If the simple annual interest was 4 percent, how many years did it take to receive the amount?

Solution:

The following information is given:

Principal (P) = $2,500.00

Balance(B) = $2,700.00

Rate (R) = 4 percent

Simple interest (I) = B − P
$$= \$2,700.00 - \$2,500.00$$
$$= \$200.00$$

Calculate the time using the following formula:

$$I = \frac{PRt}{100} \Rightarrow t = \frac{I \times 100}{PR}$$

$$t = \frac{I \times 100}{PR}$$

$$= \frac{200 \times 100}{2,500 \times 4}$$

$$= \frac{200}{100}$$

$$= 2 \text{ years}$$

So it took 2 years to receive the amount.

Example 2:

Use the simple interest formula to find the unknown quantity.

I = $840.00

P = $7,000.00

R = 4 percent

t = _?_ years

Solution:

The following information is given:

Principal (P) = $7,000.00

Interest (I) = $840.00

Interest rate (R) = 4 percent

Calculate the time using the following formula:

$$I = \frac{PRt}{100} \Rightarrow t = \frac{I \times 100}{PR}$$

$$t = \frac{I \times 100}{PR}$$

$$= \frac{840 \times 100}{7,000 \times 4}$$

$$= \frac{84}{28}$$

$$= 3 \text{ years}$$

So the time is 3 years.

Write the answer.

1. Jiten took a loan of $6,800.00 from a bank. He paid $7,616.00 to pay off the entire loan. If the simple annual interest rate was 4 percent, how many years did he take to pay off the loan?

Answer: ____ _____
unit

2. Jay borrowed $1,800.00 from his friend. He paid $2,250.00 to pay off the total amount. If the simple annual interest rate was 5 percent, how many years did he take to pay off the amount?

Answer: ____ _____
unit

Write the answer.

3. Use the simple interest formula to find the unknown quantity.

I = $400.00

P = $5,000.00

R = 2 percent

t = __?__ years

Answer: _____ _____
unit

4. Julie took a loan of $12,000.00 from a bank that gives simple annual interest of 5 percent. Lora took a loan of $10,000.00 from another bank that gives 4 percent simple annual interest. Both of them have to pay a balance of $15,000.00 to the bank. How many more years will Lora take than Julie to pay the expected balance of $15,000.00?

Answer: _____ _____
unit

5. Lisa took a loan of $20,000.00 from a bank that gives simple annual interest of 8 percent. Juhi took a loan of $16,000.00 from another bank that gives 4 percent simple annual interest. Both of them have to pay a balance of $24,000.00 to the bank. How many more years will Juhi take than Lisa to pay the expected balance of $24,000.00?

Answer: _____ _____
unit

6. Bill took a loan of $5,000.00 from a bank. He paid $5,500.00 to pay off the entire loan. If the simple annual interest rate was 10 percent, how many years did he take to pay off the loan?

Answer: _____ _____
unit

7. Use the simple interest formula to find the unknown quantity.

I = $640.00
P = $8,000.00
R = 8 percent
t = ? years

Answer: _____ _____
unit

8. Use the simple interest formula to find the unknown quantity.

I = $190.00
P = $950.00
R = 10 percent
t = ? years

Answer: _____ _____
unit

9. Kavya borrowed $10,000.00 from her friend. She paid $10,300.00 to pay off the total amount. If the simple annual interest rate was 6 percent, how many years did she take to pay off the amount?

Answer: _____ _____
unit

7.5 Find the Principal (**)

Example 1:

Anand deposited an amount in his bank account that earned $250.00 simple interest. Luke deposited an amount in his bank account that earned $325.00 simple interest. Both of them had a balance of $800.00 in their accounts after 3 years. How much more principal did Anand deposit than Luke to get the expected balance of $800.00?

Solution:

As given in the problem:

* Find the principal Anand deposited to have a balance of $800.00.

  Balance (B) = $800.00

  Simple interest (I) = $250.00

  Principal (P) = $B - I$

  $\qquad$ = $800.00 - $250.00

  $\qquad$ = $550.00

Anand deposited $550.00 to have a balance of $800.00.

* Find the principal Luke deposited to have a balance of $800.00.

  Balance (B) = $800.00

  Simple interest (I) = $325.00

  Principal (P) = $B - I$

  $\qquad$ = $800.00 - $325.00

  $\qquad$ = $475.00

Luke deposited $475.00 to have a balance of $800.00.

Difference between the principals

$\qquad$ = $550.00 - $475.00

$\qquad$ = $75.00

So Anand deposited $75.00 more than Luke to get the expected balance of $800.00.

Example 2:

Jack is investing in a mutual fund. He will be getting a simple interest rate of 4 percent per year. How much money should he invest to earn $900.00 interest in 5 years?

Solution:

The following information is given:

$\qquad$ Simple interest (I) = $900.00

$\qquad$ Time (t) = 5 years

$\qquad$ Annual interest rate (R) = 4 percent

Use these values in the simple interest formula to find P.

$$I = \frac{PRt}{100} \quad \Rightarrow \quad P = \frac{I \times 100}{Rt}$$

$$P = \frac{I \times 100}{Rt}$$

$$= \frac{900 \times 100}{5 \times 4}$$

$$= \frac{90000}{20}$$

$$= \$4,500.00$$

So Jack should invest $4,500.00.

Write the answer.

1. An investment account earned a simple interest of $340.00 in 3 years. If the balance at the end was $1,100.00, how much money was invested at the beginning?

    Answer: _____

2. Nancy deposited an amount in her bank account that earned $300.00 simple interest. Juhi deposited an amount in his bank account that earned $400.00 simple interest. Both of them want to have a balance of $1,400.00 in their accounts after 5 years. How much more money should Nancy deposit than Juhi to get the expected balance of $1,400.00?

    Answer: _____

3. Jay deposited an amount in his bank account that earned $250.00 simple interest. Sam deposited an amount in his bank account that earned $325.00 simple interest. Both of them want to have a balance of $750.00 in their accounts after 2 years. How much more money should Jay deposit than Sam to get the expected balance of $750.00?

    Answer: _____

4. John is investing in a mutual fund. He will be getting a simple interest rate of 9.5 percent per year. How much money should he invest to earn $1,710.00 interest in 3 years?

    Answer: _____

5. An investment account earned simple interest of $620.00 in 3 years. If the balance at the end was $1,500.00, how much money was invested at the beginning?

    Answer: _____

6. Julie deposited some money in her bank account. She will be getting a simple interest rate of 10 percent per year. How much money should she deposit to earn $1,400.00 interest in 2 years?

    Answer: _____

7. Bill is investing in a mutual fund. He will be getting a simple interest rate of 10 percent per year. How much money should he invest to earn $2,000.00 interest in 5 years?

    Answer: _____

7.6 Review of Chapter 7 (**)

Write or choose the letter of the answer.

1. Bob deposited $6,580.00 in a bank account that will earn a simple annual interest rate of 7.25 percent for a period of 48 months. What is the time in years?

    Answer: _____

2. What is the total interest received on an investment of $4,600.00 at an interest rate of 3 percent per year in 2 years?

    Answer: _____

3. Ada deposited $10,000.00 in her bank account. At the end of 8 years, her bank balance had gone up by $800.00. Find the simple annual interest rate.

    Answer: _____

4. Kunal is investing in a mutual fund. He will be getting a simple interest rate of 8 percent per year. How much money should he invest to earn $1,600.00 interest in 4 years?

    Answer: _____

5. Alka deposited an amount in her bank account that earned $400.00 simple interest. Ana deposited an amount in her bank account that earned $550.00 simple interest. Both of them want to have a balance of $1,000.00 in their accounts after 3 years. How much more money should Alka deposit than Ana to get the expected balance of $1,000.00?

    Answer: _____

6. Use the simple interest formula to find the unknown quantity.

    $I = \$1,260.00$
    $P = \$3,000.00$
    $R = 7$ percent
    $t = ?$ years

    Answer: ____ _____

    unit

7. When you borrow money from a bank, would you want a low or a high interest rate?

    (a) Low interest rate
    (b) High interest rate
    (c) Does not matter if the interest rate is low or high

    Answer: _____

Write or choose the letter of the answer.

8. Kate is depositing some money in her bank account. She will be getting a simple interest rate of 6 percent per year. How much money should she deposit to earn $1,800.00 interest in 2 years?

    Answer: _____

9. Use the simple interest formula $\left(I = \dfrac{PRt}{100} \right)$ to find the unknown quantity.

    I = $60.00
    P = $200.00
    R = ? percent
    t = 6 years

    Answer: _____

10. A principal amount of $6,000.00 earns simple interest of $600.00 in 4 years. What is the simple annual interest rate?

    Answer: _____

11. An investment account earned simple interest of $630.00 in 4 years. If the balance at the end was $2,700.00, how much money was invested at the beginning?

    Answer: _____

12. Lisa took a loan of $8,000.00 from a bank that gives a simple annual interest rate of 5 percent. Juhi took a loan of $10,000.00 from another bank that gives a 4 percent simple annual interest rate. Both of them have to pay a balance of $14,000.00 to the bank. How many more years will Lisa take than Juhi to pay the expected balance of $14,000.00?

    Answer: ____ _____
              unit

13. Allen invested $9,000.00 in a mutual fund. At the end of 10 years, his balance had gone up by $1,800.00. Find the simple annual interest rate.

    Answer: _____

14. Andy is investing in a municipal fund. He will be getting a simple interest rate of 11 percent per year. How much money should he invest to earn $5,500.00 interest over 10 years?

    Answer: _____

15. What is the total interest received on an investment of $10,200.00 at an interest rate of 8 percent per year in 3 years?

    Answer: _____

8. Work Problems

8.1 Concepts of Work and Time - 1

Example 1:

A group of 6 people can use 2 pounds of sugar in a week. How many people will use 6 pounds of sugar in a week?

Solution:

This problem can be solved using the following steps:

Step 1: Find the number of people that use 1 pound of sugar.

Number of people that use 2 pounds of sugar = 6

Number of people that use 1 pound of sugar = 6 ÷ 2 = 3 people

Step 2: Find the number of people that use 6 pounds of sugar.

Number of people that use 1 pound of sugar = 3

Number of people that use 6 pounds of sugar = 3 × 6 = 18 people

So 18 people will use 6 pounds of sugar in a week.

Note: For a given time, more people use more sugar, and fewer people use less sugar.

Example 2:

It takes 3 students to clean an auditorium in 6 hours. How many students will it take to clean the same auditorium in 3 hours?

Solution:

This problem can be solved using the following steps:

Step 1: Find the number of students required to clean the auditorium in 1 hour.

Number of students to clean an auditorium in 6 hours = 3

Number of students to clean the auditorium in 1 hour = 3 × 6 = 18 students

Step 2: Find the number of students required to clean the auditorium in 3 hours.

Number of students to clean the auditorium in 1 hour = 18

Number of students to clean the auditorium in 3 hours = 18 ÷ 3 = 6 students

So 6 students will take 3 hours to clean the auditorium.

Note: For a given task, more students finish the task in less time, and fewer students finish the task in more time.

Write the answer.

1. It takes 6 students to make 2 projects in a month. How many students can make 5 projects in a month?

    Answer: _____ _____
    unit

2. It takes 6 people to make 48 toy planes in a day. If they have to make 80 toy planes in the same time, how many people are needed?

    Answer: _____ _____
    unit

3. It takes 8 women to move a cart of bricks in 2 hours. How many women will it take to move the same amount of bricks in 4 hours?

    Answer: _____ _____
    unit

4. It takes 5 people to complete 3 assignments in a day. How many people will it take to complete 9 assignments in a day?

    Answer: _____ _____
    unit

5. It takes 18 students to plant a lawn in 4 hours. How many students will it take to plant the same lawn in 6 hours?

    Answer: _____ _____
    unit

6. It takes 12 cats to drink 3 liters of water in a day. How many cats will it take to drink 8 liters of water in a day?

    Answer: _____ _____
    unit

7. It takes 6 workers to dig 3 acres of garden in a day. If they have to dig 8 acres in the same time, how many workers are needed?

    Answer: _____ _____
    unit

8. It takes 5 carpenters to make a designer bed in 3 days. How many carpenters will it take to make the same bed in 5 days?

    Answer: _____ _____
    unit

8.2 Concepts of Work and Time - 2

Example 1:

It takes 2 workers 6 hours to repair a car. How many hours will it take 4 workers to repair the same car?

Solution:

This problem can be solved using the following steps:

Step 1: Find the time required for 1 worker to repair the car.

Time required for 2 workers to repair the car = 6 hours

Time required for 1 worker to repair the car
= 6 × 2
= 12 hours

Step 2: Find the time required for 4 workers to repair the car.

Time required for 1 worker to repair the car
= 12 hours

Time required for 4 workers to repair the car = 12 ÷ 4
= 3 hours

So 4 workers will take 3 hours to repair the car.

Example 2:

Mr. Rao's family can use a 3-kilogram gas cylinder in a week (7 days). How many days will it take them to use a 15-kilogram gas cylinder?

Solution:

This problem can be solved using the following steps:

Step 1: Find the number of days it takes the family to use 1 kilogram of gas.

Number of days to use 3 kg of gas = 7

Number of days to use 1 kg of gas = 7 ÷ 3

$$= \frac{7}{3} \text{ days}$$

Step 2: Find the number of days it takes the family to use 15 kilograms of gas.

Number of days to use 1 kg of gas = $\frac{7}{3}$

Number of days to use 15 kg of gas

$$= \frac{7}{\cancel{3}} \times \cancel{15}^{5}$$

$$= 7 \times 5 = 35 \text{ days}$$

So Mr. Rao's family will take 35 days to use a 15-kilogram gas cylinder.

Write the answer.

1. Raman can type 5 sentences in a minute (60 seconds). How many minutes will it take him to type 20 sentences?

2. It takes 3 robots to assemble a bike in 6 hours. If 2 robots have to assemble the bike, how many hours will it take?

Answer: ____ _____
unit

Answer: ____ _____
unit

Write the answer.

3. It takes 10 people to make some furniture in 4 days. How many days will it take 8 people to make the furniture?

Answer: _____ _____
            unit

7. It takes 2 pumps to empty a tank in 40 minutes. If 4 pumps have to empty the tank, how many minutes will they take?

Answer: _____ _____
            unit

4. It takes 3 students to finish a worksheet in 8 hours. If 4 students have to finish the worksheet, how many hours will they take?

Answer: _____ _____
            unit

8. A committee can appoint 48 members in a day (24 hours). How many hours will it take to appoint 36 members?

Answer: _____ _____
            unit

5. A company can produce 15 vehicles in a month (30 days). How many days will it take the company to produce 40 vehicles?

Answer: _____ _____
            unit

9. It takes 8 machines to wrap some gifts in a factory in 25 minutes. If 5 machines have to wrap the gifts, how many minutes will it take?

Answer: _____ _____
            unit

6. It takes 9 engineers to complete a survey in 4 hours. If 6 engineers have to complete the survey, how many hours will they take?

Answer: _____ _____
            unit

10. It takes 5 tailors 16 hours to sew some clothes. How many hours will it take 8 tailors to sew the same clothes?

Answer: _____ _____
            unit

8.3 Concepts of Work and Time - 3

Example 1:

Roni takes 8 days to design a template. How much time will she take to design three-fourths of the template?

Solution:

Time taken to design the whole template
$$= 8 \text{ days}$$

Time taken to design $\dfrac{3}{4}$ of the template

$$= \left(8 \times \dfrac{3}{4}\right) \text{ days}$$

$$= 6 \text{ days}$$

So Roni will take 6 days to design three-fourths of the template.

Example 2:

Daniel takes 4 minutes to pack 1 gift. How long will he take to pack 8 gifts?

Solution:

Time taken to pack 1 gift = 4 minutes
Time taken to pack 8 gifts = 4 × 8
$$= 32 \text{ minutes}$$

So Daniel will take 32 minutes to pack 8 gifts.

Example 3:

Angela and 2 of her friends take 8 hours to organize a party. If 3 more friends join them, how long will they take to organize the party?

Solution:

This problem can be solved using the following steps:

Step 1: Find the time required by one person to organize the party.

Angela and 2 of her friends (3 people in total) take 8 hours to organize the party.

Time taken by 3 people to organize the party = 8 hours

Time required by 1 person to organize the party = 8 × 3
$$= 24 \text{ hours}$$

Step 2: Find the time required by 6 people to organize the party.

There will be 6 people in total after 3 more friends join.

Time required by 1 person to organize the party = 24 hours

Time required by 6 people to organize the party = 24 ÷ 6
$$= 4 \text{ hours}$$

So if 3 more friends join them, it will take 4 hours to organize the party.

Write the answer.

1. A machine takes 54 minutes to make a part. How much time will it take to make two-thirds of the part?

Answer: _____ _____
unit

2. An oven takes 25 minutes to bake a cake. How long will it take to bake 5 cakes?

Answer: _____ _____
unit

Write the answer.

3. Mr. Agrawal's family is using their emergency food supply during a trip. If they eat 12 packets every day, the supply will last for 5 days. If they eat 10 packets every day, how long will the food supply last?

Answer: _____ _____
unit

4. Anita takes 8 hours to make a sweater. How much time will it take her to make one-fourth of the sweater?

Answer: _____ _____
unit

5. Mr. Harper and 4 of his coworkers take 7 hours to decorate a hotel for a wedding. If 2 more workers join them, how long will it take to decorate the hotel?

Answer: _____ _____
unit

6. Some workers take 15 days to repair a 1-kilometer-long road. How long will it take them to repair a 3-kilometer-long road?

Answer: _____ _____
unit

7. Anil takes 30 minutes to write a program. How many minutes will it take him to write 6 programs?

Answer: _____ _____
unit

8. Andrew and his family are using their emergency water tank. If they use 100 liters of water every day, the tank will last for 10 days. If they use 125 liters of water every day, how long will the tank last?

Answer: _____ _____
unit

9. Nikita and 2 of her friends take 5 hours to make some pots. If 2 more friends join them, how long will they take to make the pots?

Answer: _____ _____
unit

10. Jack takes 39 minutes to run in a race. How much time will it take him to run two-thirds of the race?

Answer: _____ _____
unit

8.4 Work Problems with Different Rates

Example 1:

Max takes 30 minutes to wash cloths. Ana takes 15 minutes to wash the same cloths. If Max and Ana wash together, how long will they take to wash the cloths?

Solution:

You can consider the whole cloths as 1 unit and solve the problem using the following steps:

Step 1: Find the amount of cloths that Max and Ana can wash in 1 minute.

- Work done by Max in 30 minutes
$$= 1 \text{ unit}$$
Work done by Max in 1 minute
$$= \left(\frac{1}{30}\right) \text{units}$$

- Work done by Ana in 15 minutes
$$= 1 \text{ unit}$$
Work done by Ana in 1 minute
$$= \left(\frac{1}{15}\right) \text{units}$$

- Work done by Max and Ana in 1 minute
$$= \frac{1}{30} + \frac{1}{15} = \frac{1}{30} + \frac{2}{30} = \frac{3}{30} \text{ units}$$

Step 2: Find the time taken to wash the whole cloths together.

Time to wash $\frac{3}{30}$ units of cloths = 1 minute

Time to wash 1 unit of cloths
$$= 1 \div \frac{3}{30} = 1 \times \frac{30}{3} = \frac{30}{3} = 10 \text{ minutes}$$

So Max and Ana will take 10 minutes to wash the cloths together.

Example 2:

There are 2 taps (tap 1 and tap 2) to fill a tank. When used individually, tap 1 can fill the tank in 2 hours and tap 2 in 3 hours. If both taps are used together, how long will it take to fill the tank? Write the answer as a fraction.

Solution:

You can consider the whole tank as 1 unit and solve the problem using the following steps:

Step 1: Find the part of the tank that can be filled by both taps in 1 hour.

- Tank filled by tap 1 in 2 hours = 1 unit
Tank filled by tap 1 in 1 hour = $\left(\frac{1}{2}\right)$ units

- Tank filled by tap 2 in 3 hours = 1 unit
Tank filled by tap 2 in 1 hour = $\left(\frac{1}{3}\right)$ units

- Tank filled by both taps in 1 hour
$$= \frac{1}{2} + \frac{1}{3} = \frac{3}{6} + \frac{2}{6} = \frac{5}{6} \text{ units}$$

Step 2: Find the time taken to fill the whole tank using tap 1 and tap 2 together.

Time to fill $\frac{5}{6}$ units of the tank = 1 hour

Time to fill 1 unit of the tank
$$= 1 \div \frac{5}{6} = 1 \times \frac{6}{5} = \frac{6}{5} \text{ hours}$$

So both tap 1 and tap 2 will take $\frac{6}{5}$ hours to fill the tank.

Write the answer.

1. Jacob takes 10 days to make a project. Elina takes 15 days to make the same project. If Jacob and Elina work together, how long will they take to make the project?

   Answer: _____ _____
   unit

2. Two students (student 1 and student 2) are watering a garden. When working individually, student 1 can water the garden in 45 minutes, and student 2 can water the garden in 30 minutes. If the students work together, how long will it take them to water the garden?

   Answer: _____ _____
   unit

3. Roshan takes 10 hours to design a building. Sofia takes 15 hours to design the same building. If Roshan and Sofia work together, how long will they take to design the building?

   Answer: _____ _____
   unit

4. Washing machines (machine 1 and machine 2) are used to clean dresses. When used individually, machine 1 can clean the dresses in 20 minutes, and machine 2 can clean the dresses in 30 minutes. If the machines are used together, how long will it take them to clean the dresses?

   Answer: _____ _____
   unit

5. Nathan takes 4 hours to make some crafts. Angela takes 12 hours to make the same crafts. If Nathan and Angela work together, how long will they take to make the crafts?

   Answer: _____ _____
   unit

6. Binny takes 20 days to eat 5 kilograms of rice. Pamela takes 60 days to eat the same amount of rice. If Binny and Pamela eat together, how long will it take them to eat 5 kilograms of rice?

   Answer: _____ _____
   unit

8.5 Review of Chapter 8

Write the answer.

1. Tina takes 10 hours to solve some problems. Monny takes 15 hours to solve the same problems. If Tina and Monny work together, how long will they take to solve the problems?

    Answer: _____ _____
    unit

2. There are 2 pipes (pipe 1 and pipe 2) used to fill a drum. When used individually, pipe 1 can fill the drum in 10 minutes, and pipe 2 can fill the drum in 15 minutes. If the pipes are used together, how long will it take to fill the drum?

    Answer: _____ _____
    unit

3. Raghab takes 9 days to make a wall. How much time will he take to make two-thirds of the wall?

    Answer: _____ _____
    unit

4. It takes 3 camels to carry a load of 1,200 kilograms in a day. How many days will it take them to carry 6,000 kilograms?

    Answer: _____ _____
    unit

5. Nelson and 2 of his coworkers can dig 6 wells in a day. If they have to dig 12 wells in the same day, how many more workers do they need?

    Answer: _____ _____
    unit

6. A machine can print 800 newspapers in a day (24 hours). How many hours will it take to print 350 newspapers?

    Answer: _____ _____
    unit

Write the answer.

7. It takes 10 worker bees to make one teaspoon of honey in a day. How many worker bees are needed to make 7 teaspoons of honey in a day?

Answer: ____ _____
                          unit

8. It takes 12 machines to make a carton of bottles in a factory in 4 hours. If 8 machines have to make the carton of bottles, how many hours will they take?

Answer: ____ _____
                          unit

9. Kile takes 50 seconds to write 20 words. Christiana takes 75 seconds to write the same number of words. If Kile and Christiana write together, how long will it take them to write 20 words?

Answer: ____ _____
                          unit

10. John and 3 of his friends take 6 hours to complete a survey. If 2 more friends join them, how long will it take to complete the survey?

Answer: ____ _____
                          unit

11. It takes 5 workers to weed a garden in 3 hours. How many workers will take 5 hours to weed the same garden?

Answer: ____ _____
                          unit

12. Nikhil and his family are using their emergency water tank during a summer storm. If they drink 50 liters of water every day, the tank will last for 9 days. If they drink 75 liters of water every day, how long will the emergency tank last?

Answer: ____ _____
                          unit

9. Mixture Problems

9.1 Mixture Problems with Solutions (**)

Example 1:

Natasha has 50 milliliters of juice. How much water does she need to add to the solution so that the mixture will be 40 percent juice?

Solution:

You can find the answer as shown below:

Step 1: Find the total amount of the solution that would be 40 percent juice.

Quantity of juice = 50 milliliters

Quantity of juice = 40 percent (of total solution)

50 milliliters = 40% × (total solution)

$$= \frac{40}{100} \times (\text{total solution})$$

$$\frac{\overset{5}{\cancel{100}}}{\underset{2}{\cancel{40}}} \times 50 = \frac{100}{40} \times \frac{40}{100} \times (\text{total solution})$$

125 milliliters = (total solution)

Step 2: Find the amount of water needed.

Amount of water needed
            = (total solution)
                – (quantity of juice)
            = 125 milliliters – 50 milliliters
            = 75 milliliters

Natasha needs to add 75 milliliters of water to the solution.

Example 2:

A bottle of 120 milliliters of kerosene has 25 percent alcohol. What is the quantity of alcohol in the bottle?

Solution:

The following information is given:

Quantity of kerosene = 120 milliliters

Quantity of alcohol = 25 percent (of quantity of kerosene)
                = 25 percent of 120 milliliters

You can find the answer as shown below:

Quantity of alcohol = 25 percent of 120 milliliters

$$= \frac{25}{100} \text{ of } 120$$

$$= \frac{25}{100} \times 120$$

$$= \frac{1}{4} \times 120$$

$$= 30 \text{ milliliters}$$

The quantity of alcohol is 30 milliliters.

Write the answer.

1. Kunal has 40 liters of milk. How much water does he need to add to the solution so that the mixture will be 80 percent milk?

4. Two drums (drums A and B) each contain 250 liters of gas. Drum A is 15 percent kerosene, and drum B is 30 percent kerosene. How much more kerosene does drum B have than drum A?

Answer: ____ _____
unit

Answer: ____ _____
unit

2. Two bottles (bottles 1 and 2) each have 150 milliliters of solution in them. Bottle 1 is 50 percent water, and bottle 2 is 30 percent water. How much more water does bottle 1 have than bottle 2?

5. A 450-milliliter solution has 60 percent oil. What is the quantity of oil in the solution?

Answer: ____ _____
unit

Answer: ____ _____
unit

6. Arun has 20 liters of sprite in his shop. How much orange soda does he need to add to the solution so that the mixture will be 20 percent sprite?

3. A 500-milliliter bottle contains 75 percent liquid chemical. What is the amount of liquid chemical in the bottle?

Answer: ____ _____
unit

Answer: ____ _____
unit

9.2 Mixture Problems with Objects (**)

Example 1:

A bag of 20 kilograms of rice is 30 percent wheat. How much wheat is in the bag?

Solution:

You can find the answer as shown below.

Quantity of rice = 20 kilograms

Quantity of wheat

$= 30$ percent (of quantity of rice)

$= 30$ percent of 20

$= \dfrac{30}{100}$ of 20

$= \dfrac{30}{100} \times 20$

$= \dfrac{30}{5} \times 1$

$= 6$ kilograms of wheat

There is 6 kilograms of wheat in the bag.

Example 2:

Mr. Baker has 21 armchairs in his farmhouse. How many folding chairs does he need to add so that the mixture will be 50 percent armchairs?

Solution:

You can find the answer as shown below.

Step 1: Find the total number of chairs in a mixture that is 50 percent armchairs.

Number of armchairs = 21

Number of armchairs = 50 percent (of total chairs)

$21 = \dfrac{50}{100} \times$ (total chairs)

$\dfrac{\overset{2}{\cancel{100}}}{\cancel{50}} \times 21 = \dfrac{100}{50} \times \dfrac{50}{100} \times$ (total chairs)

$2 \times 21 =$ (total chairs)

$42 =$ (total chairs)

Step 2: Find the number of folding chairs that need to be added.

Number of folding chairs needed

$=$ (total chairs)

$\qquad -$ (number of armchairs)

$= 42 - 21 = 21$

Mr. Baker needs to add 21 folding chairs.

Write the answer.

1. A shelf of 25 books is 20 percent storybooks. How many storybooks are there on that shelf?

2. A bag of 30 kilograms of fruit is 45 percent bananas. What quantity of bananas is in the bag?

Answer: _____ _____
unit

Answer: _____ _____
unit

Name _____

Write the answer.

3. A farmer has 80 cows. How many buffaloes does he need to add to the mixture so that the mixture will be 40 percent cows?

Answer: ____ _____
_{unit}

4. Two aquariums (1 and 2) have 30 fish each. Aquarium 1 has 40 percent red fish, and aquarium 2 has 60 percent red fish. How many more red fish are there in aquarium 2 than in aquarium 1?

Answer: ____ _____

unit

5. Mr. Ford has 18 digital paintings. How many oil paintings does he need to add to the mixture so that the mixture will be 90 percent digital paintings?

Answer: ____ _____

unit

6. A company has 210 male employees. How many female employees does it need to hire so the company will be 70 percent male employees?

Answer: ____ _____

unit

7. Two apartments (1 and 2) have 60 vehicles each. Apartment 1 has 35 percent motorcycles, and apartment 2 has 45 percent motorcycles. How many more motorcycles are there in apartment 2 than in apartment 1?

Answer: ____ _____
unit

8. Peter has 39 T-shirts. How many collared shirts does he need to add to the mixture so that the mixture will be 75 percent T-shirts?

Answer: ____ _____
unit

9. Two lunch packs (A and B) have 12 items each. Lunch pack A has 50 percent burgers, and lunch pack B has 25 percent burgers. How many more burgers are there in lunch pack A than in lunch pack B?

Answer: ____ _____
unit

9.3 Mixture Problems with Cost (**)

Example 1:

Bag A contains 10 books that cost $9.00 per book. Bag B contains 5 books that cost $6.00 per book. If you mix the bags together, what will the cost per book in the mixture be?

Solution:

You can find the answer using the following steps:

- Find the total cost of the books in bag A.
  Number of books in bag A = 10
  Cost per book = $9.00
  Total cost of books in bag A
  $$= 10 \times 9 = \$90.00$$

- Find the total cost of the books in bag B.
  Number of books in bag B = 5
  Cost per book = $6.00
  Total cost of books in bag B
  $$= 5 \times 6 = \$30.00$$

- Total books in the mixture
  = (number of books in bag A)
     + (number of books in bag B)
  $$= 10 + 5 = 15$$

- Total cost of the books
  = (total cost of books in bag A)
     + (total cost of books in bag B)
  $$= \$90.00 + \$30.00 = \$120.00$$

- Cost per book in the mixture
  = (total cost of the books)
     ÷ (total books in the mixture)
  $$= \$120.00 \div 15 = \$8.00$$

So the cost per book in the mixture will be $8.00.

Example 2:

A pack contains 12 chocolates that cost $2.25 per chocolate. What is the total cost of chocolates in the pack?

Solution:

You can find the answer as shown below.

Number of chocolates in the pack = 12

Cost per chocolate = $2.25

Total cost of the pack
   = (number of chocolates in the pack)
                × (cost per chocolate)
   $$= 12 \times \$2.25$$
   $$= \$27.00$$

The total cost of chocolates in the pack is $27.00.

Write the answer.

1. Room 1 has 20 computers that cost $90.00 per computer, and room 2 has 30 computers that cost $80.00 per computer. If you mix the computers in both rooms, what will the cost per computer in the mixture be?

   Answer: _____

2. Amit has 8 toys that cost $4.40 per toy. What is the total cost of the toys?

   Answer: _____

3. Table 1 contains 12 dinner plates that cost $5.00 per plate, and table 2 contains 18 dinner plates that cost $5.00 per plate. If you mix the plates from both tables, what will the cost per dinner plate in the mixture be?

   Answer: _____

4. Kiran has 6 greeting cards that cost $5.35 per card. What is the total cost of the greeting cards?

   Answer: _____

5. A bucket contains 11 bottles that cost $8.99 per bottle. What is the total cost of bottles in the bucket?

   Answer: _____

6. A box contains 15 toothpaste tubes that cost $1.75 per tube. What is the total cost of tubes in the box?

   Answer: _____

7. Shop A contains 20 crystal balls that cost $30.00 per ball, and shop B contains 30 crystal balls that cost $40.00 per ball. If you mix the crystal balls from both shops, what will the cost per crystal ball in the mixture be?

   Answer: _____

8. Box A has 20 necklaces that cost $10.50 per necklace, and box B has 10 necklaces that cost $6.00 per necklace. If you mix the necklaces in both boxes, what will the cost per necklace in the mixture be?

   Answer: _____

9.4 Filling Containers with One or Two Pipes (*)

Example 1:

A water pipe can fill a pool in 44 minutes. How long will the pipe take to fill one-fourth of the pool?

Solution:

You can consider the whole pool to be 1 and find the answer as shown below.

Time the water pipe takes to fill the whole (1) pool = 44 minutes

Time the water pipe takes to

fill $\frac{1}{4}$ of the pool = $\cancel{44} \times \frac{1}{\cancel{4}}$

$= 11$ minutes

It will take 11 minutes to fill one-fourth of the pool.

Example 2:

Tap 1 can fill a drum in 10 minutes. Tap 2 can fill the same drum in 15 minutes. If both taps are opened at the same time when the drum is empty, how long will it take to fill the drum?

Solution:

You can find the answer using the following steps.

- Find the part of the drum that can be filled by tap 1 in 1 minute.

  Part of the drum filled by tap 1 in 10 minutes = 1

  Part of the drum filled by tap 1 in 1 minute = $\frac{1}{10}$

- Find the part of the drum that can be filled by tap 2 in 1 minute.

  Part of the drum filled by tap 2 in 15 minutes = 1

  Part of the drum filled by tap 2 in 1 minute = $\frac{1}{15}$

- Find the part of the drum that can be filled by both tap 1 and tap 2 in 1 minute.

  Part of the drum filled by both taps in 1 minute = $\frac{1}{10} + \frac{1}{15} = \frac{3}{30} + \frac{2}{30} = \frac{5}{30}$

- Find the time taken to fill the drum by both taps together.

  Time taken to fill $\frac{5}{30}$ of drum by both taps = 1 minute

  Time taken to fill 1 (whole drum) by both taps = $1 \div \frac{5}{30} = 1 \times \frac{30}{5} = 6$ minutes

Write the answer.

1. A water pipe can fill an empty bucket in 60 seconds. How long will the pipe take to fill three-fourths of the bucket?

   Answer: _____ _____
   unit

2. Tap A can fill a water tank in 3 hours. Tap B can fill the same tank in 6 hours. If both taps are opened at the same time when the tank is empty, how long will it take to fill the tank?

   Answer: _____ _____
   unit

3. Pipe A can fill a drum in 6 minutes. Pipe B can fill the same drum in 12 minutes. If both the pipes are opened at the same time when the drum is empty, how long will it take to fill the drum?

   Answer: _____ _____
   unit

4. A water tap can fill an empty drum in 45 minutes. How long will the tap take to fill four-fifths of the drum?

   Answer: _____ _____
   unit

5. Pipe 1 can fill a swimming pool in 12 hours. Pipe 2 can fill the same pool in 24 hours. If both the pipes are opened at the same time when the pool is empty, how long will it take to fill the pool?

   Answer: _____ _____
   unit

6. An electric pump can fill an empty container in 24 minutes. How long will the pump take to fill two-thirds of the container?

   Answer: _____ _____
   unit

9.5 Emptying Containers with One or Two Pipes (*)

Example 1:

Tap A can empty a full tank in 20 hours, and tap B can empty the same tank in 30 hours. If both taps are started at the same time while the tank is full, how long will it take to empty the tank?

Solution:

You can find the answer as shown below:

• Find the fraction of the tank that can be emptied by tap A in 1 hour.

Fraction of tank emptied by tap A in 20 hours = 1

Fraction of tank emptied by tap A in 1 hour = $\dfrac{1}{20}$

• Find the fraction of the tank that can be emptied by tap B in 1 hour.

Fraction of tank emptied by tap B in 30 hours = 1

Fraction of tank emptied by tap B in 1 hour = $\dfrac{1}{30}$

• Find the fraction of the tank that can be emptied by both taps in 1 hour.

Fraction of tank emptied by both taps in 1 hour = $\dfrac{1}{20} + \dfrac{1}{30} = \dfrac{3}{60} + \dfrac{2}{60}$

$= \dfrac{5}{60}$

• Find the time taken by both taps to empty the tank.

Time taken to empty $\dfrac{5}{60}$ of the tank by both taps = 1 hour

Time taken by both taps to empty the tank

$= 1 \div \dfrac{5}{60} = 1 \times \dfrac{60}{5}$

$= 12$ hours

Example 2:

A water pipe can empty a full container in 6 hours. How long will the pipe take to empty one-third of the container?

Solution:

You can consider the whole container to be 1 and find the answer as shown below.

Time taken by water pipe to empty the whole (1) container = 6 hours

Time taken by water pipe to empty $\dfrac{1}{3}$ of the container = $\overset{2}{\cancel{6}} \times \dfrac{1}{\cancel{3}}$

$= 2 \times 1$

$= 2$ hours

It will take 2 hours to empty one-third of the container.

Write the answer.

1. A nozzle (nozzle 1) can empty an oil tank in 10 hours. Another nozzle (nozzle 2) can empty the oil tank in 15 hours. If both nozzles are opened at the same time while the tank is full, how long will it take to empty the oil tank?

   Answer: _____ _____
                              unit

2. A pipe can empty a full container in 8 hours. How long will the pipe take to empty two-fourths of the container?

   Answer: _____ _____
                              unit

3. Pipe A can empty a pool in 15 hours. Pipe B can empty the same pool in 30 hours. If both the pipes are opened at the same time while the pool is full, how long will it take to empty the pool?

   Answer: _____ _____
                              unit

4. A pump can empty a full drum in 35 minutes. How long will the pump take to empty three-fifths of the drum?

   Answer: _____ _____
                              unit

5. Tap A can empty a water tank in 24 minutes. Tap B can empty the same tank in 48 minutes. If both the taps are opened at the same time while the tank is full, how long will it take to empty the water tank?

   Answer: _____ _____
                              unit

6. A nozzle can empty a full tank in 54 minutes. How long will the nozzle take to empty five-sixths of the tank?

   Answer: _____ _____
                              unit

7. A tap can empty a full tank in 6 hours. How long will the pump take to empty two-thirds of the tank?

   Answer: _____ _____
                              unit

9.6 Review of Chapter 9 (**)

Write the answer.

1. Two schools (A and B) each have 450 students. School A is 40 percent boys, and school B is 60 percent boys. How many more boys does school B have than school A?

    Answer: _____ _____
               unit

2. A water tap can fill an empty container in 18 hours. How long will the tap take to fill two-sixths of the container?

    Answer: _____ _____
               unit

3. A jar contains 25 cookies that cost $0.80 per cookie. What is the total cost of cookies in the jar?

    Answer: _____ _____
               unit

4. A 750-milliliter bottle is 20 percent alcohol. What is the quantity of alcohol in the bottle?

    Answer: _____ _____
               unit

5. Shop A has 26 umbrellas that cost $20.00 per umbrella, and shop B has 24 umbrellas that cost $30.00 per umbrella. If you mix the umbrellas in both shops, what will the cost per umbrella in the mixture be?

    Answer: _____ _____
               unit

6. Kevin has 6 fountain pens. How many ballpoint pens does he need to add to the solution so the mixture will be 10 percent fountain pens?

    Answer: _____ _____
               unit

7. A pipe can fill a bucket in 28 seconds. How long will it take the pipe to fill three-fourths of the bucket?

    Answer: _____ _____
               unit

Write the answer.

8. Shop 1 has 90 milk bottles that cost $12.00 per bottle, and shop 2 has 10 milk bottles that cost $22.00 per bottle. If you mix the milk bottles in both shops, what will the cost per milk bottle in the mixture be?

Answer: _____ _____
unit

11. Two sports clubs (clubs A and B) each have 40 players in them. Club A is 55 percent football players, and club B is 70 percent football players. How many more football players does club B have than club A?

Answer: _____ _____
unit

9. Nozzle A can fill an oil tank in 9 hours. Nozzle B can fill the same tank in 18 hours. If both nozzles are opened at the same time when the tank is empty, how long will it take to fill the oil tank?

Answer: _____ _____
unit

12. A pump can empty a full tank in 20 hours. How long will the pump take to empty two-fourths of the tank?

Answer: _____ _____
unit

13. Sarah has 16 toy planes. How many toy horses does she need to add to the mixture so the mixture will be 80 percent toy planes?

Answer: _____ _____
unit

10. Pipe A can empty a tank in 40 minutes. Pipe B can empty the same tank in 60 minutes. If both pipes are opened at the same time while the tank is full, how long will it take to empty the tank?

Answer: _____ _____
unit

14. A pack contains 6 ice-cream cones that cost $4.40 per ice-cream cone. What is the total cost of ice-cream cones in the pack?

Answer: _____ _____
unit

Quiz

1. What operation will you use for the key word *double*?
   - (a) Division
   - (b) Multiplication
   - (c) None of the above

   Answer: _____

2. Nathan got $28.00 from his father. If he spent three-fourth of the money on snacks, how much did he spend on snacks?

   Answer: _____

3. Mark has 150 red and white roses in his shop. The number of red roses is 25 more than the number of white roses. Use the variable N to represent the number of white roses. Which equation represents the given problem?
   - (a) $N - 25 = 150$
   - (b) $2N + 25 = 150$
   - (c) $2N - 25 = 150$
   - (d) $N + 25 = 250$

   Answer: _____

4. Pamela spent $60.00 at a shopping mall. She bought 2 tops which cost the same amount and a watch for $32.00. What was the cost of each top?

   Answer: _____

5. Kelvin reads 60 pages in 3 hours. How many pages will he read in 9 hours?

   Answer: _____

6. It takes 4 people 3 hours to decorate for a party. How long will it take for 2 people to do the same decoration?

   Answer: _____

7. A number has 2 decimal places. It is greater than 24 and less than 25. The tenths-place digit is 3 more than the hundredths-place digit. If the hundredths-place digit is 5, what is the decimal number?

   Answer: _____

8. I am the largest decimal number with two decimal places. If I am smaller than 39 and ends with the digit 7, what number am I?

   Answer: _____

111

9. Ava is 7 years younger than George. The sum of their ages is 23. How old is Ava?

Answer: _____

10. Nancy is currently 3 times as old as Maria. Six years ago, the difference between their ages was 18. What is Nancy's current age?

Answer: _____

11. A tiger ran 153 miles in 3 hours. What was the average speed in miles per hour?

Answer: _____

12. Mr. White has to travel 160 miles by car to visit his construction site. He completed the first 100 miles at a speed of 50 miles per hour. Due to poor road condition, he travelled the remaining distance at a speed of 30 miles per hour. What was the average speed during the entire trip?

Answer: _____

13. What is the total interest received on an investment of $975.00 at an interest rate of 4 percent per year in 2 years?

Answer: _____

14. Kelvin invested $8,000.00 in a mutual fund. At the end of 3 years, his balance had gone up by $1,200.00. Find the simple annual interest rate.

Answer: _____

15. It takes 6 workers to mow a garden in 4 hours. How many workers will take 8 hours to mow the same garden?

Answer: _____

16. Kile takes 20 minutes to water plants in a garden. Adriana takes 30 minutes to do the same work. If Kile and Adriana work together, how long will it take them to water the garden?

Answer: _____

17. A box contains 12 pizzas that cost $2.50 per pizza. What is the total cost of pizzas in the box?

Answer: _____

18. Tap A can empty a tank in 30 minutes. Tap B can empty the same tank in 45 minutes. If both taps are opened at the same time, how long will it take to empty the full tank?

Answer: _____

www.ingramcontent.com/pod-product-compliance
Lightning Source LLC
Chambersburg PA
CBHW081154180526
45170CB00006B/2073